Architectural Design and Construction

Instructor's Manual

DEPARTMENT OF HEALTH AND HUMAN SERVICES
Centers for Disease Control and Prevention
National Institute for Occupational Safety and Health

Disclaimer

Mention of any company or product does not constitute endorsement by NIOSH. In addition, citations to Web sites external to NIOSH do not constitute NIOSH endorsement of the sponsoring organizations or their programs or products. Furthermore, NIOSH is not responsible for the content of these Web sites.

Ordering Information

This document is in the public domain and may be freely copied or reprinted. To receive NIOSH documents or other information about occupational safety and health topics, contact NIOSH at

Telephone: 1–800–CDC–INFO (1–800–232–4636)
TTY: 1–888–232–6348
Web site: www.cdc.gov/info

or visit the NIOSH Web site at www.cdc.gov/niosh

For a monthly update on news at NIOSH, subscribe to *NIOSH eNews* by visiting www.cdc.gov/niosh/eNews.

DHHS (NIOSH) Publication No. 2013–133

March 2013

Safer • Healthier • People™

Please direct questions about these instructional materials to the National Institute for Occupational Safety and Health (NIOSH):

Telephone: (513) 533–8302
E-mail: preventionthroughdesign@cdc.gov

Foreword

A strategic goal of the Prevention through Design (PtD) Plan for the National Initiative is for designers, engineers, machinery and equipment manufacturers, health and safety (H&S) professionals, business leaders, and workers to understand the PtD concept. Further, they are to apply these skills and this knowledge to the design and redesign of new and existing facilities, processes, equipment, tools, and organization of work. In accordance with the PtD Plan, this module has been developed for use by educators to disseminate the PtD concept and practice within the undergraduate engineering curricula.

John Howard, M.D.
Director, National Institute for
   Occupational Safety and Health
Centers for Disease Control and Prevention

Contents

Acknowledgments

Authors:

Michael Behm, Ph.D.
Cory Boughton

The authors thank the following for their reviews:

NIOSH Internal Reviewers

Pamela E. Heckel, Ph.D., P.E.
Donna S. Heidel, M.S., C.I.H.
Thomas J. Lentz, Ph.D., M.P.H.
Rick Niemeier, Ph.D.
Andrea Okun, Ph.D.
Paul Schulte, Ph.D.
Pietra Check, M.P.H.
John A. Decker, Ph.D.
Matt Gillen, M.S., C.I.H.
Roger Rosa, Ph.D.

Peer and Stakeholder Reviewers

Michael J. Buono, A.I.A., LEED A.P.
Joe Fradella, Ph.D.
Kihong Ku, Ph.D.
Matthew Marshall, Ph.D.
Gopal Menon, P.E.
Virginia L. Russell, F.A.S.L.A., LEED A.P.
James Platner, Ph.D.
Georgi Popov, Ph.D.
Deborah Young-Corbett, Ph.D., C.I.H., C.S.P., C.H.M.M.

Introduction

This Instructor's Manual is part of a broad-based multi-stakeholder initiative, Prevention through Design (PtD). This module has been developed for use by educators to disseminate the PtD concept and practice within the undergraduate engineering curricula. Prevention through Design anticipates and minimizes occupational safety and health hazards and risks[*] at the design phase of products,[†] considering workers through the entire life cycle, from the construction workers to the users, the maintenance staff, and, finally, the demolition team. The engineering profession has long recognized the importance of preventing occupational safety and health problems by designing out hazards. Industry leaders want to reduce costs by preventing negative safety and health consequences of poor designs. Thus, owners, designers, and trade contractors all have an interest in the final design.

This manual is one of four PtD education modules to increase awareness of construction hazards. The modules support undergraduate courses in civil and construction engineering. The four modules cover the following:

1. Reinforced concrete design
2. Mechanical–electrical systems
3. Structural steel design
4. **Architectural design and construction.**

This manual is specific to a PowerPoint slide deck related to Module 4, Architectural design and construction. It contains learning objectives, slide-by-slide lecture notes, case studies, test questions, and references. It is assumed that the users are experienced professors/lecturers in schools of engineering/architecture. As such, the manual does not provide specifics on *how* the materials should be presented. However, background insights are included on most of the slides for the instructor's consideration.

Numerous examples of inadequate design and catastrophic failures can be found on the Internet. If time permits, have the students seek, share, and analyze appropriate and inadequate designs. The PtD Web site is located at www.cdc.gov/niosh/topics/ptd. The National Institute for Occupational Safety and Health (NIOSH) Fatality Assessment and Control Evaluation (FACE) Reports can be found at www.cdc.gov/niosh/face/. Occupational Safety and Health Administration (OSHA) Fatal Facts are available at www.setonresourcecenter.com/MSDS_Hazcom/FatalFacts/index.htm.

---

[*] A "hazard" is anything with the potential to do harm. A "risk" is the likelihood of potential harm from that hazard being realized.

[†] The term *products* under the Prevention through Design umbrella pertains to structures, work premises, tools, manufacturing plants, equipment, machinery, substances, work methods, and systems of work.

Learning Objectives and Overview

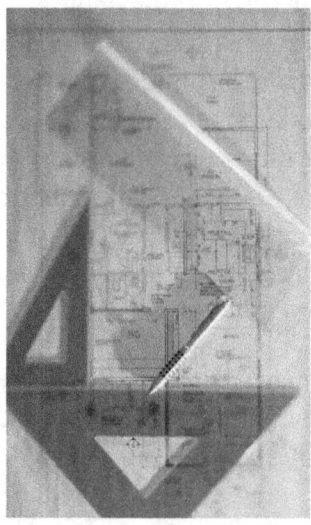

Photo courtesy of Thinkstock

Architectural Design and Construction
EDUCATION MODULE

Developed by Michael Behm , Ph.D.
Cory Boughton
East Carolina University

NOTES TO INSTRUCTORS

This module presents safe-design considerations pertaining to architectural design and construction. It contains specific examples of common workplace hazards related to construction and illustrates ways design can make a difference. There are several case studies to facilitate class discussions. One section of slides presents the Prevention through Design (PtD) concept, another set summarizes architectural design principles, and a third set illustrates applications of the PtD concept to real-world construction scenarios.

This education module is intended to facilitate incorporation of the PtD concept into your architectural design course. You may wish to supplement the information presented in this module and may assign projects, class presentations, or homework as time permits. Sections may be presented independently of the whole. Presentation times are approximate, based on our presentation experience.

To activate the features embedded in some slides, please "enable content," make this a "trusted document," and view the slides in "slide show" mode. To show the presentation file in slideshow mode, press F5. Each slide is accompanied by speaker notes that you can read aloud while the slide is projected on the screen. The audience does not see the speaker notes. When you click on "Use Presenter View" on the Slide Show tab, your monitor displays the speaker notes but the projected image does not.

Thank you for using this module. To report problems or to make suggestions, please contact the National Institute for Occupational Safety and Health (NIOSH):

Telephone: (513) 533–8302
E-mail: preventionthroughdesign@cdc.gov

SOURCE
Photo courtesy of Thinkstock

 Guide for Instructors

Slides	Slide numbers	Approx. minutes
Introduction to Prevention through Design	5–28	45
Site Planning	29–34	10
Excavation	35–40	10
Building Elements	41–65	50
General Considerations	66–68	5
Building Decommissioning	69–71	5
Recap	72–73	5
References and Other Sources	74–88	—

NOTES

The first two slides of the presentation provide acknowledgments and general information. Learning objectives are delineated on Slide 3. Slide 4 contains the Overview. Slides 5 through 28 introduce the PtD concept and can be covered in approximately 45 minutes. The topic of slides 29 through 34 is site planning. Slides 35 through 40 present the hazards associated with excavation. Slides 41 through 65 provide specific examples of Prevention through Design opportunities for various building elements. Lifting and inhalation hazards are presented on slides 67 through 68. PtD also applies to building renovation and decommissioning; see slides 69 through 71. A summary is contained on slides 72 and 73. References are provided on slides 74 through 88. Additional time may be required to discuss the case studies.

Learning Objectives

- Explain the Prevention through Design (PtD) concept.

- List reasons why project owners may wish to incorporate PtD in their projects.

- Identify workplace hazards and risks associated with design decisions and recommend design alternatives to alleviate or lessen those risks.

NOTES

After completing this education module, you should be able to do the following:

- Explain the PtD concept
- Describe motivations, barriers, and enablers for implementing PtD in projects
- List three reasons why PtD improves business value.

Overview

- PtD concept

- Site planning

- Excavation

- Building elements

- General considerations

- Decommissioning

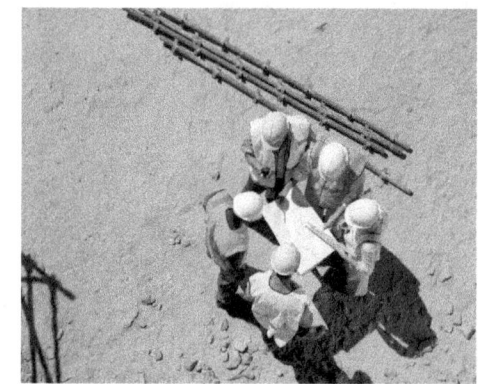

Photo courtesy of Thinkstock

NOTES

This is an overview of the PtD topics covered in this module. Many of you are not familiar with PtD, so we spend a few minutes discussing what the concept is. Next we summarize the safety concepts pertaining to site planning and excavation. Then we discuss specific building elements and general safety considerations. Finally, we look at specific hazards associated with decommissioning a building.

Introduction to
Prevention through Design (PtD)

Introduction to Prevention through Design

EDUCATION MODULE

NOTES

Let's start by introducing PtD.

Occupational Safety and Health

- Occupational Safety and Health Administration (OSHA)
  www.osha.gov
    - Part of the Department of Labor
    - Assures safe and healthful workplaces
    - Sets and enforces standards
    - Provides training, outreach, education, and assistance
    - State regulations possibly more stringent

- National Institute for Occupational Safety and
  Health (NIOSH) www.cdc.gov/niosh
    - Part of the Department of Health and Human Services, Centers
      for Disease Control and Prevention
    - Conducts research and makes recommendations for the
      prevention of work-related injury and illness

NOTES

All employers, including structural design firms, are required by law to provide their employees with a safe work environment and training to recognize hazards that may be present. They also must provide equipment or other means to minimize or manage the hazards.

Designers historically have not been familiar with the federal Occupational Safety and Health Act (OSH Act) standards because they were rarely exposed to construction jobsite hazards. However, with the increasing roles that designers are playing on worksites, such as being part of a design-build team, it is becoming increasingly important that they receive construction safety training, including information about federal and state construction safety standards.

The Occupational Safety & Health Act of 1970, Public Law 91-596 (OSH Act) [29 USC* 1900], was passed on December 29, 1970, "To assure safe and healthful working conditions for working men and women; by authorizing enforcement of the standards developed under the Act; by assisting and encouraging the States in their efforts to assure safe and healthful working conditions; by providing for research, information, education, and training in the field of occupational safety and health; and for other purposes." The construction industry standards

*United States Code. See USC in Sources.

enforced by the Occupational Safety and Health Administration (OSHA) are found in Title 29 Part 1926 of the Code of Federal Regulations [29 CFR 1926].

The National Institute for Occupational Safety and Health (NIOSH) is part of the Department of Health and Human Services, Centers for Disease Control and Prevention. The National Occupational Research Agenda (NORA) is a partnership program to stimulate innovative research and improved workplace practices. Unveiled in 1996, NORA has become a research framework for NIOSH and the nation. Diverse parties collaborate to identify the most critical issues in workplace safety and health. Partners, then, work together to develop goals and objectives for addressing these needs. Participation in NORA is broad, including stakeholders from universities, large and small businesses, professional societies, government agencies, and worker organizations. NIOSH and its partners have formed ten NORA Sector Councils: Agriculture, Forestry & Fishing; Construction; Healthcare & Social Assistance; Manufacturing; Mining; Oil and Gas Extraction; Public Safety; Other Services; Transportation, Warehousing & Utilities; and Wholesale and Retail Trade. The mission of the NIOSH research program for the Construction sector is to eliminate occupational diseases, injuries, and fatalities among individuals working in these industries through a focused program of research and prevention.

SOURCES

CFR. Code of Federal Regulations. Washington, DC: U.S. Government Printing Office, Office of the Federal Register.

NIOSH FACE reports [www.cdc.gov/niosh/face/]

Fatal Facts Accident Reports Index [www.setonresourcecenter.com/MSDS_Hazcom/FatalFacts/index.htm]

OSHA home page [www.osha.gov/]

USC. United States Code. Washington, DC: U.S. Government Printing Office.

Construction Hazards

 **Construction Hazards**

- Cuts

- Electrocution

- Falls

- Falling objects

- Heat/cold stress

- Musculoskeletal disease

- Tripping

[BLS 2006; Lipscomb et al. 2006]

Graphic courtesy of OSHA

NOTES

A construction worksite by its nature involves numerous potential hazards. A portion of the work is directly affected by weather. Workers interact with heavy equipment and materials at elevated heights, in below-ground excavations, and in multiple awkward positions. The composition of the site workforce changes over the project. Work may be done autonomously or in coordination with others. The construction worksite is unforgiving to poor planning and operational errors.

For these reasons, pre-job construction-phase planning is used as a best practice to systematically address potential hazards. Project-specific worker safety orientations prior to site work also play an important role. PtD practices, by systematically looking further upstream at design-related potential hazards, extend these pre-job measures. PtD can help identify potential hazards so that they can be eliminated, reduced, or communicated to contractors for pre-job planning.

Every hazard that can be addressed should be addressed. Falling can cause serious injury. Boilermakers, pipe-fitters, and iron workers can experience career-ending musculoskeletal injuries by lifting heavy loads or working in a cramped position. Anyone can be seriously injured by a falling object. Whether a structural member or a simple wrench, a falling object can be deadly. Anyone can trip, but the elevated height and proximity to dangerous equipment increase the risk of injury on a construction site. Some accidents are caused by poor lighting and/or sunlight glare. Common injuries due to spatial misperception include hitting your head or cutting yourself on sharp corners. Hot summer and cold winter days can affect worker health. Personal protective equipment (PPE), such as hardhats, gloves, ear protection, and safety glasses, is required for a reason! Not every hazard on a construction worksite can be "designed out," but many significant ones can be minimized during the design phase.

SOURCES

BLS [2006]. Injuries, illnesses, and fatalities in construction, 2004. By Meyer SW, Pegula SM. Washington, DC: U.S. Department of Labor, Bureau of Labor Statistics, Office of Safety, Health, and Working Conditions.

Lipscomb HJ, Glazner JE, Bondy J, Guarini K, Lezotte D [2006]. Injuries from slips and trips in construction. Appl Ergonomics 37(3):267–274.

OSHA [ND]. Fatal Facts Accident Reports Index [foreman electrocuted]. Accident summary no. 17 [www.setonresourcecenter.com/MSDS_Hazcom/FatalFacts/index.htm].

Graphic courtesy of OSHA

Fatal Facts Accident Report courtesy of OSHA

ACCIDENT REPORT
FATAL FACTS

ACCIDENT SUMMARY No. 17

Accident Type:	Electrocution
Weather Conditions:	Sunny, Clear
Type of Operation:	Steel Erection
Size of Work Crew:	3
Collective Bargaining	No
Competent Safety Monitor on Site:	Yes - Victim
Safety and Health Program in Effect:	No
Was the Worksite Inspected Regularly:	Yes
Training and Education Provided:	No
Employee Job Title:	Steel Erector Foreman
Age & Sex:	43-Male
Experience at this Type of Work:	4 months
Time on Project:	4 Hours

BRIEF DESCRIPTION OF ACCIDENT

Employees were moving a steel canopy structure using a "boom crane" truck. The boom cable made contact with a 7200 volt electrical power distribution line electrocuting the operator of the crane; he was the foreman at the site.

INSPECTION RESULTS

As a result of its investigation. OSHA issued citations for four serious violations of its construction standards dealing with training, protective equipment, and working too close to power lines.

OSHA's construction safety standards include several requirements which, If they had been followed here. might have prevented this fatality.

ACCIDENT PREVENTION RECOMMENDATIONS

1. Develop and maintain a safety and health program to provide guidance for safe operations (29 CFR 1926.20(b)(1)).
2. Instruct each employee on how to recognize and avoid unsafe conditions which apply to the work and work areas (29 CFR 1926.21(b)(2))
3. If high voltage lines are not de-energized, visibly grounded, or protected by insulating barriers, equipment operators must maintain a minimum distance of 10 feet between their equipment and the electrical distribution or transmission lines (29 CFR 1926.550(a)(15)(i)).

SOURCES OF HELP

- Ground Fault Protection on Construction Sites (OSHA 3007) which describes OSHA requirements for electrical safety at construction sites.

- Construction Safety and Health Standards (OSHA 2207) which contains all OSHA job safety and health rules and regulations (1926 and 1910) covering construction
- OSHA Safety and Health Training Guidelines for Construction (available from the National Technical Information Service - Order No PB-239312/AS) comprised of a set of 15 guidelines to help construction employees establish a training program in the safe use of equipment, tools, and machinery on the job
- OSHA-funded free onsite consultation services Consult your telephone directory for the number of your local OSHA area or regional office for further assistance and advice (listed under the US Labor Department or under the state government section where states administer their own OSH programs).

NOTE: The case here described was selected as being representative of fatalities caused by improper work practices. No special emphasis or priority is implied nor is the case necessarily a recent occurrence. The legal aspects of the incident have been resolved, and the case is now closed.

Construction Accidents

 **Construction Accidents in the United States**

Construction is one of the most hazardous occupations. This industry accounts for

- 8% of the U.S. workforce, but 20% of fatalities

- About 1,100 deaths annually

- About 170,000 serious injuries annually

[CPWR 2008]

Photo courtesy of Thinkstock

Architecture

CDC Workplace NIOSH

NOTES

As many of us know, construction is one of the most dangerous industries for workers. In the United States, construction typically accounts for 170,000 serious injuries and 1,100 deaths each year. The fatality rate is disproportionally high for the size of the construction workforce. Twenty percent of all collapses during construction are the result of structural design errors. Statistics like these do not tell the whole story. Behind every serious injury, there is a real story of an individual who suffered serious pain and may never fully recover. Behind every fatality, there are spouses, children, and parents who grieve every day for their loss. We all recognize that safety is a vital component of an inherently dangerous business. All of us—including architects and engineers—must do what we can to reduce the risk of injuries on our projects.

SOURCES

CPWR [2008]. The construction chart book. 4th ed. Silver Spring, MD: Center for Construction Research and Training.

New York State Department of Health [2007]. A plumber dies after the collapse of a trench wall. Case report 07NY033 [www.cdc.gov/niosh/face/pdfs/07NY033.pdf].

OSHA [ND]. Fatal facts Accident Reports Index [foreman electrocuted]. Accident summary no. 19 [www.setonresourcecenter.com/MSDS_Hazcom/FatalFacts/index.htm].

Photo courtesy of Thinkstock

Face report courtesy of NY State Department of Health.

Fatal Facts Accident Report courtesy of OSHA

NEW YORK

state department of

HEALTH

FATALITY ASSESSMENT AND CONTROL EVALUATION

A Plumber Dies After the
Collapse of a Trench Wall
Case Report: 07NY033

SUMMARY

In May 2007, a 46 year old self-employed plumbing contractor (the victim) died when the unprotected trench he was working in collapsed. The victim was an independent plumber subcontracted to install a sewer line connection to the sewer main, part of a general contractor project to install a new sanitary sewer for an existing single family residence.

At approximately 12:30 PM on the day of the incident, the workers on site observed the victim walking back toward the residence for parts as they initiated their lunch break. When the victim did not come for his lunch or answer his cell phone, the general contractor and workers starting searching for the victim. The excavation contractor observed that a portion of the trench had collapsed where the victim was installing a sewer tap. The victim was found trapped in the trench under a large slab of asphalt, rock and soil. Three workers immediately climbed down the side of the trench to try to assist the victim. One of the workers called 911 on his cell phone. Police and emergency medical services (EMS) arrived on site within minutes. The EMS members entered the unprotected trench but could not revive the victim. The county trench rescue team recovered the victim's body at approximately seven feet below grade and lifted him from the ditch four hours after the incident. He was pronounced dead at the site. More than 50 rescue workers were involved in the recovery.

New York State Fatality Assessment and Control Evaluation (NY FACE) investigators concluded that, to help prevent similar occurrences, employers and independent contractors should:
- **Require that all employees, subcontractors, and site workers working in trenches five feet or more in depth are protected from cave-ins by an adequate protection system.**
- **Require that a competent person conducts daily inspections of the excavations, adjacent areas, and protective systems and takes appropriate measures necessary to protect workers.**
- **Require that all employees and subcontractors have been properly trained in the recognition of the hazards associated with excavation and trenching. In addition, the general contractor (GC) should be responsible for the collection and review of training records and require that all workers employed on the site have received the requisite training to meet all applicable standards and regulations for the scope of work being performed.**
- **Require that on a multi-employer work site, the GC should be responsible for the coordination of all high hazard work activities such as excavation and trenching.**

- **Require that all employees are protected from exposure to electrical hazards in a trench.**

Additionally,

- **Employers of law enforcement and EMS personnel should develop trench rescue procedures and should require that their employees are trained to understand that they are not to enter an unprotected trench during an emergency rescue operation.**
- **Local governing bodies and codes enforcement officers should receive additional training to upgrade their knowledge and awareness of high hazard work, including excavation and trenching. This skills upgrade should be provided to both new and existing codes enforcement officers.**
- **Local governing bodies and codes enforcement officers should consider requiring building permit applicants to certify that they will follow written excavation and trenching plans in accordance with applicable standards and regulations, for any projects involving excavation and trenching work, before the building permits can be approved.**

INTRODUCTION

In May, 2007, a 46 year old self-employed plumbing contractor died when the trench he was working in collapsed at a residential construction site. Approximately 8000 pounds of broken asphalt, rock and dirt fell atop the victim, fatally crushing him as he was installing a sewer tap to a town sewer main. The New York State Fatality Assessment and Control Evaluation (NY FACE) program learned about the incident from a newspaper article the following day. The Occupational Safety and Health Administration (OSHA) investigated the incident along with the county sheriff's office. The NY FACE staff met and reviewed the case information with the OSHA compliance officer. This report was developed based upon the information provided by OSHA, the county sheriff's department, and the county coroner's medical and toxicological reports.

The general contractor (GC) on the residential construction site had been hired by the homeowners to complete a project that included the installation of a new sanitary sewer connection for an existing single family residence. The GC was the owner and sole employee of his company, which had been in business for many years. The GC directed the work of two subcontractors on the work site to complete the installation of the residential sewer line.

- One subcontractor was an excavating company that had been in business for approximately four years. The owner of this company hired two workers to assist him with the excavation of the trench.
- The second subcontractor was the victim, a self-employed licensed plumber who had over twenty years of experience with a variety of construction projects, including the installation of sewer lines. The victim did not have any previous work relationship with either the GC or the excavation subcontractor.

The OSHA investigation report indicated that the GC and the subcontractor did not have health and safety programs. A formal health and safety plan had not been established to identify the hazards of the excavation project and the actions to be taken to remediate them. The GC, subcontractors and the subcontractors' employees did not have hazard recognition training or safety training on the fundamentals of excavation and trenching. None of the workers on the site were knowledgeable on excavation and trenching safety standards and applicable regulations and they did not understand the hazards and dangers associated with working in a trench. A competent person was not present to conduct initial and ongoing inspections of the excavation project, identify potential health and safety hazards such as possible cave-in, and oversee the use of adequate protection systems and work practices.

INVESTIGATION

The GC was hired to replace a crushed sewer line that ran under the driveway of an existing single family residence. Rather than dig up the driveway to replace the old line, which was thought to be more costly and time-consuming, the GC decided to run a new line. All required town permits had been obtained and the local codes enforcement requirements for one-call system notification of the excavation and underground utility location mark-outs had been completed. The work had been scheduled to be completed in one day (Friday), but the excavation subcontractor lost time due to hitting a water line and encountering very rocky soil during the excavation. The project had to be extended to two days (Friday and Monday). The town water and sewer inspector visited the work site on Friday, observed the digging of the trench which began at the residence, and halted the digging of the trench at the edge of the property to avoid having an open trench in the road and consequent road closure over a weekend. Excavation company workers had been observed in the trench spotting and hand digging.

On Monday, the day of the incident, the excavating subcontractor initiated excavation from the edge of the road to the sewer main in the roadway. An employee witness of the excavating company stated that the victim was directing excavation work while in the trench and hand digging to expose the sewer main once the excavator came close to the location. OSHA findings indicated that tools were uncovered in the trench in the area of the trench wall collapse, including a shovel, pick ax, hammer drill and drill bits, consistent with the scenario of the victim being in the ditch, hand digging to locate the sewer main. The town water and sewer inspector also visited the work site on Monday. He determined that the victim did not have the correct parts to complete the sewer connection, advised him of the correct parts, and indicated that he would return later in the day to re-inspect and photograph the completed sewer tap in order to allow the excavating subcontractor to run the pipe back to the house, backfill the excavation and reopen the road.

The GC left the work site to purchase the correct parts, while the excavation continued. The dimensions of the final trench were approximately 55 feet in length, 3 feet to 8 feet in depth, and 30 inches to 128 inches in width (see Figure 1). It was shaped like a "T." The gravity sewer main that the victim was connecting to was located at a depth of 7 feet 4 inches (7' 4") below grade at the east (E) end of the top of the "T." Installation of new sewer pipe from the residence had been initiated and some of the trench had already been backfilled. The length of the trench from the top of the "T" to the location of the newly installed sewer pipe was 35 feet 11 inches (35'11") at the time of the incident. Soil analysis results, conducted after the incident, indicated a granular, sandy gravel Type C soil (OSHA Excavation Standard) that contained large cobbles and boulders, the least stable soil type.

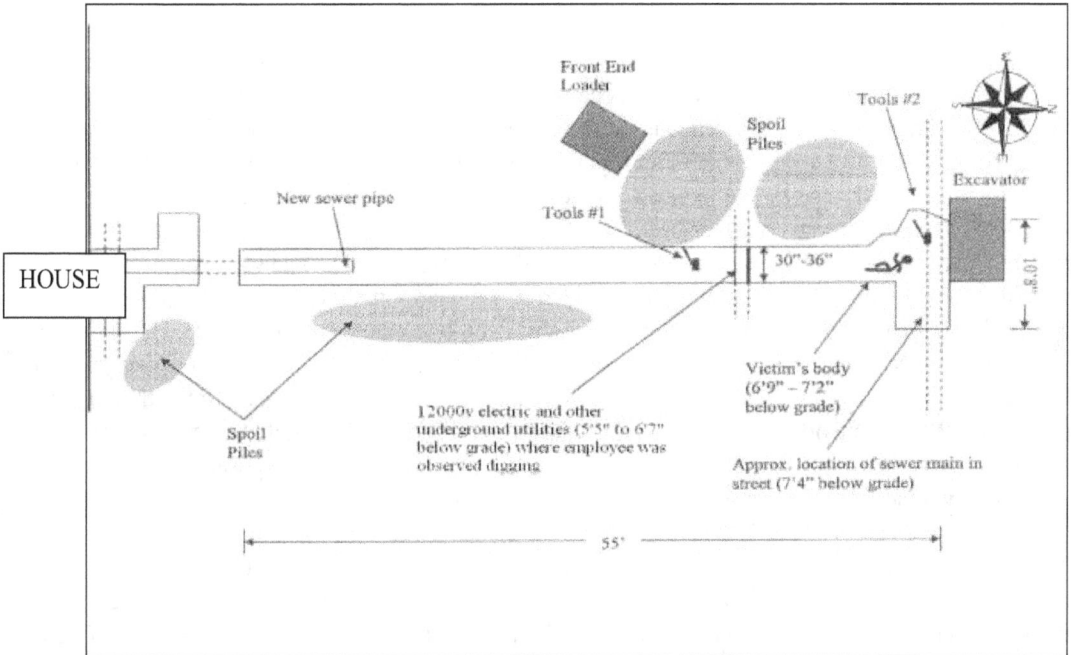

Figure 1: Schematic of the excavation and the incident site (courtesy of OSHA)

The faces of the trench were vertical. No shoring or benching was used. Large cobbles and boulders and loose rock/dirt were visible on the face of the excavation and were not removed or supported. The pavement above the E and W faces of the excavation had been undermined during excavation activities and no support system was utilized to protect employees from a possible collapse. Pieces of road pavement and asphalt had been undermined during excavation activities in the road in the proximity of the sewer main at the top of the "T." These areas were in plain view and did not have additional support. On the W side of the excavation, loose boulders, rock and debris in spoils piles were located less than two feet from the edge of the trench. (Figure 2) The excavator was positioned adjacent to the N end of the trench, where undermined areas were in plain sight. The N end of the trench, where the victim was installing the sewer tap, also lacked an access ladder or other safe means of entry/egress.

Figure 2: View of the west wall of the excavation south of the "T."
Note the boulders and loose rock/dirt on the excavation face as well as the location of the spoils pile within 2 feet of the edge of the trench. (courtesy of OSHA)

The GC returned just before 12 noon with the correct parts and handed them to the victim. The GC left the site in order to purchase lunch for the workers, including the victim. At this same time, the victim called the town water and sewer inspector, informed him that he had located the sewer main, had all the correct parts, and was ready to connect. The town inspector informed the victim that someone from the town would be out after lunch to inspect and photograph the sewer tap. According to the town inspector, a sewer tap to a sewer main is a simple job that would take about 20 minutes to complete. The GC returned with lunch at 12:30PM. The workers, with the exception of the victim, took a break for lunch at a location near the front end loader (Figure 1). The workers saw the victim walking in the trench in the direction of the residence and heard him say that he was "looking for a splitter for a three-way." By 1:00 PM the victim still had not come for his lunch. The GC called the victim on his cell phone and looked for him in his van behind the house. The other workers joined in the search. The excavating subcontractor observed that a portion of the west side of the trench had collapsed. When the workers approached the excavation, they found the victim trapped in the trench under a large slab of asphalt, rock, and soil, with only the back of his head exposed. Three workers climbed down the side of the trench to try to assist the victim.

The workers removed the dirt from around his head, lifted his head, and tried to clear his airway. They checked for a pulse, but found none. One of the workers then called 911 from his cell phone. The workers attempted to move the slab of asphalt without success. Within minutes, the police arrived, followed by EMS at approximately 1:08 PM. The EMS personnel entered the unprotected trench but were unable to revive the victim. Volunteer firefighters from multiple fire departments and a special trench rescue team responded, the latter team having been created by the county after the deaths of two workers in a construction trench collapse 10 years earlier. A wooden safety box was built by the trench rescue team and efforts began to free the victim from entrapment by chipping the asphalt slab into pieces. Using a system of ropes and pulleys, the rescue team lifted the victim from the ditch at 4:25 PM. His body had been recovered at about 7' below grade. The county coroner pronounced him dead at 4:35 PM. Approximately 50 rescuers responded to the 911 call.

The OSHA investigation resulted in findings that the trench section that collapsed was a triangular shaped area at the northwest corner of the excavation, approximately 5 feet 1 inch (5' 1") in length, 4 feet (4') wide, and 6-7 feet (6-7') deep. Multiple hazards were present, but had not been identified and remediated. The W side of the excavation collapsed and pieces of asphalt paving and rock fatally crushed the victim while he was making the sewer tap (Figures 3 and 4).

The hazards of the unprotected trench exposed additional people to the excavation collapse as the GC, the excavation company workers and EMS personnel entered the trench to attempt a rescue of the victim. In addition to the trench hazards, no precautions had been taken to prevent exposure to the underground electrical and utility lines. The town inspector had noted that a young employee of the excavation company was "manually hand digging with shovel and pick ax "within a few inches of the buried electrical lines." This is consistent with OSHA findings that indicated attempts had been made to cut the rock in the face of the trench at the location of the underground utilities. A demo saw, hammer drill and cordless reciprocating saw used to cut rocks and pavement were found within inches of the 12,000 volt underground electrical line. Several other utilities were also exposed in this location at the edge of the road (Figure #1, Tools #1). EMS personnel also entered the trench when power was still connected to the utilities in the trench.

Figure 3: Location of collapse.
Note spoils piles and equipment located less
2 feet from the edge of the trench
(courtesy of OSHA)

Figure 4: Area of trench collapse
Note the large boulders hanging from the than
excavation faces and undermined areas on the
edge of the trench (courtesy of OSHA)

RECOMMENDATIONS/DISCUSSION

Recommendation #1: *Employers and independent contractors should require that all employees, subcontractors and site workers working in trenches five feet or more in depth are protected from cave-ins by an adequate protection system.*

Discussion: Employers and contractors should require that all employees working in trenches five feet deep or more are protected from cave-ins by an adequate protection system appropriate to the conditions of the trench, including sloping techniques or support systems such as shoring or trench boxes (OSHA 29CFR 1926.652). Sloping involves positioning the soil away from an excavation trench at an angle that would prevent the soil from caving into the trench. Even in shallow trenches less than five feet in depth, the possibility of accidents still exists. Trenches five feet deep or less should also be protected if a competent person identifies a cave-in potential. Trench protection systems are available to all employers and independent contractors, even as rental equipment. Employers should also require that all pieces of excavated pavement, asphalt, dirt, rock, boulders, and debris as well as excavation equipment are located in spoils piles or positions that are at least two feet from the edge of the excavated trench. Where a two foot setback is not possible, spoils may need to be hauled to another location. In this incident, sloping would not have been an appropriate protection system, due to the composition of the soil. Employers and contractors should consult tables located in the appendices of the OSHA Excavation Standard that detail the protection required based upon the soil type and environmental conditions present at a work site. Employers and contractors can also consult with manufacturers of protective systems to obtain detailed guidance for the appropriate use of protection systems.

Trenches should be kept open only for the minimum amount of time needed. Hinze and Bren (1997) observed that the risk of a collapse in an unprotected trench increases the longer a trench is open. They propose that after a trench is dug, the apparent cohesion of trench walls may begin to relax after only four hours, contributing to increasingly unstable walls in an unprotected trench. In this incident, a 45 feet length of the trench had been excavated and was left open for more than two days. The trench section where the incident occurred was dug at approximately 8:30 AM on the day of the incident. Hand digging and incorrect parts resulted in additional delays in making the sewer tap to the main. The trench collapse occurred approximately four hours later, between 12:30 PM and 1:00 PM.

The key to preventing a trench accident is not to enter an unprotected trench. When the walls of a trench collapse or cave in, the results are entrapment or struck-by incidents to anyone caught inside, accidents which can occur in seconds. Many workers in a trench are in a kneeling or squatting position that results in little opportunity for an escape. Victims do not need to be completely covered in soil. Even with partial covering, enough pressure is created for mechanical asphyxia in which the weight of the dirt and soil compresses the chest. One cubic yard of soil has an average weight of 2500 pounds (Figure 4), but can vary due to the composition and moisture content.

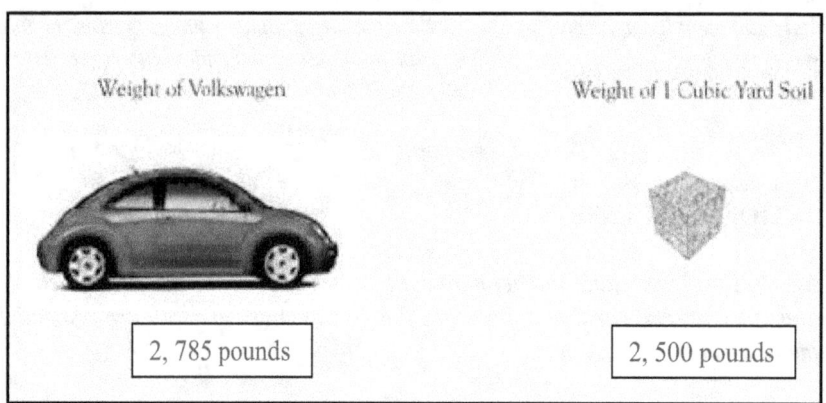

Figure 5: Weight of one cubic yard of soil (courtesy of "Weights of Building Materials, Agricultural Commodities, and Floor Loads for Buildings" standard reference)

Recommendation #2: *Employers and independent contractors should require that a competent person conducts daily inspections of the excavations, adjacent areas, and protective systems and takes appropriate measures necessary to protect workers.*

Discussion: Employers and independent contractors are responsible for complying with the OSHA Excavation Standard requirements to designate a competent person on site for excavation and trenching projects to make daily inspections of excavations, the adjacent areas, and protective systems (OSHA 29CFR 1926.651). A competent person is defined as someone who is capable of identifying existing and predictable hazards in the surroundings and working conditions that are dangerous to employees and who has the authorization to take prompt corrective measures to eliminate them. They should inspect the trenches daily, as needed throughout the work shift, and as conditions change (for example, heavy rainfall or increased traffic vibrations). These inspections should be conducted before worker entry, to ensure that there is no evidence of a possible cave-in, failure of a protective system, hazardous conditions such as spoils piles or equipment location, or hazardous atmosphere.

In particular, competent persons are required by OSHA to complete a competent person training curriculum, which could be an OSHA training program or an equivalent safety or trade organization training. The competent person needs be knowledgeable on the hazards associated with excavation and trenching, as well as the causes of injuries and the safe work practices and specific protective actions needed. Competent persons must also be experienced in excavation and trenching with a minimum of hands-on training in a demonstration trench or in a field component. The competent person needs to know the key points of the OSHA Excavation Standard, including the excavation standards and appendices, checklists, soils analysis and the components of a daily trenching inspection.

Having a competent person is a particularly acute problem among contracting companies that employ fewer than 10 workers. Of the National Institute for Occupational Safety and Health (NIOSH) FACE cases related to excavation and trenching, 88% were non-union companies with less than 10 workers. These small companies are not members of trade associations and are the least likely to employ trench safety protections and to have an adequately trained competent person or an excavation crew.

In this incident, no competent person was hired by the GC to conduct initial and ongoing inspections of the trench. The GC, excavating contractor, and excavation company employees did not possess an understanding of the hazards associated with excavation and trenching operations or a knowledge of the requirements of the OSHA Excavation Standard. No one on-site was qualified to function as the competent person.

Recommendation #3*: Employers and independent contractors should require that all employees and subcontractors have been properly trained in the recognition of the hazards associated with excavation and trenching. On a multi-employer work site, the GC should be responsible for the collection and review of training records and require that all workers employed on the site have received the requisite training to meet all applicable standards and regulations for the scope of work being performed.*

Discussion*:* Excavation and trenching is one of the most hazardous construction operations. Even with a competent person on site, workers in excavation and trenching operations are also in need of health and safety training, including basic hazard recognition and prevention. Workers should be able to identify the specific hazards associated with excavation and trenching, the reasons for using protective equipment and how to work in a trench safely. Workers should be trained not to enter an unprotected trench, even in a rescue attempt, since they place themselves at risk of becoming injured or killed. If necessary, projects should be delayed until training requirements are met and training records are provided.

In this case, the general contractor, excavation subcontractor, and excavation company employees did not demonstrate adequate knowledge of safe work practices in excavation and trenching. The limited training in proper excavation technique as well as inadequate hazard recognition and prevention training were critical to the failure to properly assess the hazards present and protect the trench.

Recommendation #4: *Employers and independent contractors should require that on a multi-employer work site, the GC should be responsible for the coordination of all high hazard work activities such as excavation and trenching.*

Discussion: The GC is responsible and accountable for the safety of all employees, subcontractors and workers on the site. Health and safety plans should be in place to formally address the hazards that may be encountered, including written plans to manage these hazards and protect the safety of all workers on the site.

In this incident, the GC did coordinate the work activities of the subcontractors and workers on the job, but health and safety plans were not addressed. The management of excavation and trenching hazards was left to a subcontractor who was not a competent person, knowledgeable or trained in the requirements of the OSHA Excavation Standard.

Recommendation #5: *Employers of law enforcement and EMS personnel should develop trench rescue procedures and should require that their employees are trained to understand that they are not to enter an unprotected trench during an emergency rescue operation.*

Discussion: Employers of law enforcement and EMS personnel should develop a formal safety procedure for emergency rescue in an unprotected trench. Entering an unprotected trench after a cave-in or collapse could place would-be rescuers in danger. Rescue is a delicate and slow operation requiring knowledge of the behavior of unstable soil, necessary to prevent further injury to the victim or the rescuers. The added weight and vibrations can also contribute to an increased susceptibility to further collapse. Many rescuers precipitate second and third stage trench cave-ins and have become victims themselves. In this incident EMS personnel entered the unprotected trench in an attempt to rescue the victim, exposing themselves to an excavation collapse hazard.

Emergency rescue workers, such as law enforcement officials and EMS personnel, should receive specialized training in how to rescue workers who may be trapped in utility trenches, and should not put themselves in danger by entering an unprotected trench. In this incident, a specialized rescue team was called in to respond to the emergency. The rescue workers had special equipment for trench rescues and building collapses and had undergone specialized training in the area of trench/building collapse emergencies. They immediately constructed a wooden safety box in the trench with a system of ropes and pulleys before entering the trench to free the victim. National Fire Protection Association (NFPA) 1670, Chapter 11 details the requirements for rescue operations after a trench cave-in occurs.

Recommendation #6: *Local governing bodies and codes enforcement officers should receive additional training to upgrade their knowledge and awareness of high hazard work, including excavation and trenching. This skills upgrade should be provided to both new and existing codes enforcement officers.*

Discussion: This recommendation may create a mechanism of observation and oversight by the codes enforcement officers who are likely to encounter small employers and independent contractors during their work. The officers could inform the employers and contractors of potential hazards, provide fact sheets that highlight the key requirements for the excavation and trenching standards, and check some of the basics of the trenching project such as depth of the trench, protection of the trench and identification of the competent person. In addition, they could advise employers and contractors to contact safety experts to learn about and implement trench safety. This may be an effective accident prevention strategy, reaching the thousands of untrained and unprepared small employers and independent contractors with awareness and guidance, the very workers who represent the major group of fatalities in New York State.

In this incident, the town water and sewer inspector observed workers in the unprotected trench serving as spotters, observed a worker hand digging within a few feet of a live buried electrical utility, and observed the victim spotting in the unprotected trench for the excavating subcontractor while attempting to locate the sewer main. If the above recommendation was in place, with a trained and knowledgeable officer, at a minimum the excavation work may have been halted and entry into an unprotected trench may have been prohibited.

Recommendation #7: *Local governing bodies and codes enforcement officers should consider requiring building permit applicants to certify that they will follow written excavation and trenching plans in accordance with applicable standards and regulations, for any projects involving excavation and trenching work, before the building permits can be approved.*

Discussion: Local governing bodies may consider revising building permits to require building permit applicants to certify that they will follow written plans for any projects involving excavation and trenching. Statements on the permit applications would be added to indicate that the employer/independent contractor agrees to accept and abide by all standards and regulations governing the excavation and trenching work, not just local governing body codes and ordinances. If construction companies and independent contractors were required to provide written documentation of how the high hazard work of excavation and trenching will be performed safely as part of the building permit application process, it may prompt the employers and contractors to plan ahead, formally assess the hazards, seek assistance in developing the required safety and injury prevention program, and implement the necessary injury prevention measures. No work should be initiated unless these requirements are met after review and approval. These changes may help to prevent trench related fatalities in NYS.

Recommendation #8: *Employers and independent contractors should require that all employees are protected from exposure to electrical hazards in a trench.*

Discussion: Utilities to the single family residence were located underground in the trench near the edge of the road. Workers were observed using power and hand tools within inches of live 12,000 volt lines. This did not contribute to the fatality, but did present another potential hazard to workers in the excavation and trenching project and to the rescue workers. Performing cutting work next to hot utility lines could have resulted in additional serious injuries and death from electrocution. The company performed the utility mark-out as required by local codes but did not contact the utility company to turn off the power as required, when they realized the need to hand cut large rocks and boulders in the trench. The power was not shut off to these lines until after the incident, when workers returned to complete the work.

Key words: Trench, collapse, cave-in, trenching, excavation, trench protection systems, entrapment, spoils piles

REFERENCES:

1. Associated General Contractors of America Safety Training for the Focus Four. *Hazards in Construction.* Retrieved February 8, 2011 from http://www.agc.org/cs/career_development/safety_training/focus_four_locations

2. CDC/NIOSH. *NIOSH Safety and Health Topic: Trenching and Excavation.* Retrieved on February 8, 2011 from http://www.cdc.gov/niosh/topics/trenching/

3. CDC/NIOSH. MMWR. 2004. *Occupational Fatalities During Trenching and Excavation Work - United States, 1992-2001*. Morbidity and Mortality Weekly Report, 53(15):311-314. Retrieved February 8, 2011 from www.cdc.gov/mmwr/preview/mmwrhtml/mm5315a2.htm

4. CDC/NIOSH. Alert: July 1985. *Preventing Deaths and Injuries from Excavation Cave-ins*. retrieved February 8, 2011 from http://www.cdc.gov/niosh/85-110.html

5. CDC/NIOSH. Fatality Assessment and Control Evaluation (FACE) investigation reports. Retrieved February 8, 2011 from www.cdc.gov/niosh/face

6. Center to Protect Workers' Rights (CPWR). Plog, Barbara et al. March, 2006. *Barriers to Trench Safety: Strategies to Prevent Trenching-Related Injuries and Deaths.* Retrieved February 8, 2011 from www.elcosh.org.

7. Commonwealth of Massachusetts. Executive Office of Labor and Workforce Development. *Trenching Hazard Alert for Public Works Employers and Employees in Massachusetts.* Bulletin 407, 11/2007, p1-4.

8. Deatherage, J.H., et al. 2004 *Neglecting Safety Precautions may lead to trenching fatalities*. American Journal of Industrial Medicine, 45(6):522-7.

9. EC&M online. June, 2009. *Danger Uncovered.* Beck, Ireland. Retrieved February 8, 2011 from http://ecmweb.com/construction/electrical-trench-safety-20090601/

10. Encyclopedia of Occupational Health and Safety. 4th Edition. *Chapter 93: Construction Trenching* by Jack Mickle. *Types of Projects and Their Associated Hazards* by Jeffrey Hinkman. Retrieved February 8, 2011 from
        http://www.elcosh.org/en/document/296/d000279/encyclopedia-of-occupational-safety-%2526-health-%253A-chapter-93-construction.html

11. Executive Safety Update. The Monthly News Bulletin of the Construction Safety Center, Vol. 17, Issue 3, September, 2009

12. Hinze, J.W. and K. Bren. 1997. *The causes of trenching-related fatalities*. Construction Congress V: Managing Engineered Construction in Expanding Global Markets. Proceedings of the Congress, sponsored by the American Society of Civil Engineers (ASCE), 131(4): 494-500.

13. Irizarry, J. et al: 2002 *Analysis of Safety Issues in Trenching Operation*. 10th Annual Symposium on Construction Innovation and Global Competitiveness, September 9-13, 2002. Retrieved February 8, 2011 from Construction Safety Alliance site: http://engineering.purdue.edu/CSA/publications/trenching03

14. Job Health and Safety Quarterly. Fall, 2009. *Trenching is a Dangerous and Dirty Business.* Retrieved February 8, 2011 from http://www.elcosh.org/en/document/161/d000168/trenching-is-a-dangerous-and-dirty-business.html

15. Miami-Dade County. *Trench Safety Act Compliance Statement, FM5238 Rev. (12-00).* Retrieved February 8, 2011 from http://facilities.dadeschools.net/form_pdfs/5238.pdf

16. New York City Department of Buildings. *Excavation and Trench Safety Guidelines* by Dan Eschenasy. www.NYC.gov/buildings. Retrieved February 8, 2011 from http://www.elcosh.org/en/document/161/d000168/trenching-is-a-dangerous-and-dirty-business.html

17. OSHA. *Working Safely in Trenches Safety Tips.* Retrieved February 8, 2011 from http://www.osha.gov/Publications/trench/trench_safety_tips_card.html

18. OSHA. *29CFR1926.650 subpart p. Excavations: scope, application and definitions.* Retrieved February 8, 2011 from http://www.osha.gov/pls/oshaweb/owadisp.show_document?p_id=10774&p_table=STANDARDS

19. OSHA. *29CFR1926.651 subpart p. Excavations: specific excavation requirements.* Retrieved February 8, 2011 from http://www.osha.gov/pls/oshaweb/owadisp.show_document?p_table=STANDARDS&p_id=10775

20. OSHA. *29CFR1926.652 subpart p. Excavations: requirements for protective systems.* Retrieved February 8, 2011 from http://www.osha.gov/pls/oshaweb/owadisp.show_document?p_table=STANDARDS&p_id=10776

21. OSHA. *OSHA Technical Manual SECTION V: CHAPTER 2 EXCAVATIONS: HAZARD RECOGNITON IN TRENCHING AND SHORING.* Retrieved February 8, 2011 from http://www.osha.gov/dts/osta/otm/otm_v/otm_v_2.html

22. OSHA. *OSHA's Construction e-tool.* Retrieved February 8, 2011 from http://www.osha.gov/SLTC/etools/construction/trenching/mainpage.html

The New York State Fatality Assessment and Control Evaluation (NY FACE) program is one of many workplace health and safety programs administered by the New York State Department of Health (NYSDOH). It is a research program designed to identify and study fatal occupational injuries. Under a cooperative agreement with the National Institute for Occupational Safety and Health (NIOSH), the NY FACE program collects information on occupational fatalities in New York State (excluding New York City) and targets specific types of fatalities for evaluation. NY FACE investigators evaluate information from multiple sources and summarize findings in narrative reports that include recommendations for preventing similar events in the future. These recommendations are distributed to employers, workers, and other organizations interested in promoting workplace safety. The NY FACE does not determine fault or legal liability associated with a fatal incident. Names of employers, victims and/or witnesses are not included in written investigative reports or other databases to protect the confidentiality of those who voluntarily participate in the program.

Additional information regarding the NY FACE program can be obtained from:
New York State Department of Health FACE Program
Bureau of Occupational Health
Flanigan Square, Room 230

1-518-402-7900
www.nyhealth.gov/nysdoh/face/face.htm

ACCIDENT REPORT
FATAL FACTS

ACCIDENT SUMMARY No. 59

Accident Type:	Struck by Falling Wall	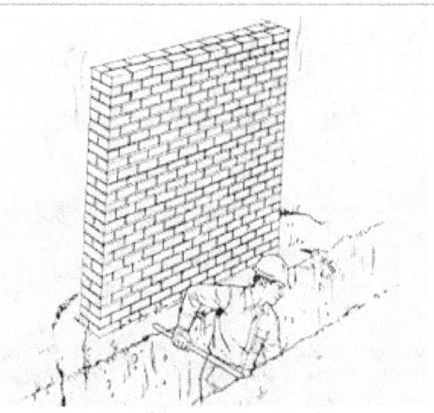
Weather Conditions:	Clear/Wet Soil	
Type of Operation:	Trenching	
Size of Work Crew:	2	
Competent Safety Monitor on Site:	No	
Safety and Health Program in Effect:	Inadeqaute	
Was the Worksite Inspected Regularly:	No, short duration	
Training and Education Provided:	Some	
Employee Job Title:	Laborer	
Age & Sex:	27-Male	
Experience at this Type of Work:	1 Year	
Time on Project:	1 Day	

BRIEF DESCRIPTION OF ACCIDENT

An employee was in the process of locating an underground water line. A trench had been dug approximately 4 feet deep along side a brick wall 7 feet high and 5 feet long. The brick wall collapsed onto the victim who was standing in the trench. The injuries were fatal.

INSPECTION RESULTS

As a result of its investigation, OSHA issued citations for violation of the standard.

ACCIDENT PREVENTION RECOMMENDATIONS

The contractor should not permit employees to excavate below the level of the base of foundation footings when walls are unpinned [29 CFR 1926.651(i)(1)]

SOURCES OF HELP

- **OSHA 2202 Construction Industry Digest** ¯ includes all OSHA construction standards and those general industry standards that apply to construction. Order No. 029-016-00151-4, ($2.25). Available from the Superintendent of Documents, Government Printing Office, Washington DC 20402-9325, phone (202) 512-1800. Make checks payable to Superintendent of Documents. For phone orders, Visa® or MasterCard®.
- **OSHA 2254 Training Requirements in OSHA Standards and Training Guidelines** ¯ includes all OSHA construction standards and those general industry standards that apply to construction. Order No. 029-016-00160-3, ($6.00). Available from the Superintendent of Documents, Government Printing Office, Washington DC 20402-9325, phone (202) 512-1800. Make checks payable to Superintendent of Documents. For phone orders, Visa® or MasterCard®.
- **OSHA Safety and Health Guidelines for Construction** (Available from the National Information Service, 5285 Port Royal Road, Springfield, VA 22161; (703) 605-6000 or (800) 553-6847; Order No. PB-239-312/AS, $27). Guidelines to helpconstruction employers establish a training program in the safe use of equipment, tools, and machinery on the job.

- For information on OSHA-funded free consultation services call the nearest OSHA area office listed in telephone directories under U.S. Labor Department or under the state government section where states administer their own OSHA programs.
- Courses in construction safety are offered by the OSHA Training Institute, 1555 Times Drive, Des Plaines, IL 60018, 708/297-4810.
- OSHA Safety and Health Training Guidelines for Construction (Available from the National Technical Information Service, 5285 Port Royal Road, Springfield, VA 22161; 703/487-4650; Order No. PB-239-312/AS): guidelines to help construction employers establish a training program in the safe use of equipment, tools, and machinery on the Job.

NOTE: The case here described was selected as being representative of fatalities caused by improper work practices. No special emphasis or priority is implied nor is the case necessarily a recent occurrence. The legal aspects of the incident have been resolved, and the case is now closed.

Scaffolding Accidents

Design as a Risk Factor: Australian Study, 2000–2002

- Main finding: design contributes significantly to work-related serious injury

- 37% of workplace fatalities are due to design-related issues

- In another 14% of fatalities, design-related issues may have played a role

[Driscoll et al. 2008]

Photo courtesy of Thinkstock

NOTES

Several studies around the world have demonstrated that design can directly affect the safety of a construction site or process. The Australian government investigated the design-related root causes of their work-related fatalities. Seventy-seven (37%) of the 210 identified workplace fatalities definitely or probably had design-related issues involved. In another 29 fatalities (14%), the circumstances suggested that design issues were involved. The most common scenarios involved problems with rollover protective structures and/or associated seat belts; inadequate guarding; lack of residual current devices; inadequate fall protection; failed hydraulic lifting systems in vehicles and mobile equipment; and inadequate protection mechanisms on mobile plants and vehicles.

These fatal incidents might have been prevented if the hazards that caused them had been considered during the design phase.

SOURCES

Driscoll TR, Harrison JE, Bradley C, Newson RS [2008]. The role of design issues in work-related fatal injury in Australia. J Safety Res 39(2):209–14 [Epub 2008:Mar 13; PubMed index for MEDLINE: 18454972].

NIOSH Fatality Assessment and Control Evaluation (FACE) Program [1983]. Fatal incident summary report: scaffold collapse involving a painter. FACE 8306 [www.cdc.gov/niosh/face/In-house/full8306.html].

Photo courtesy of Thinkstock

FACE report courtesy of NIOSH

Fatal Incident Summary Report: Scaffold Collapse Involving a Painter

INTRODUCTION

The National Institute for Occupational Safety and Health (NIOSH), Division of Safety Research (DSR), is currently conducting the Fatal Accident Circumstances and Epidemiology (FACE) Study. By scientifically collecting data from a sample of similar fatal accidents, this study will identify and rank factors which increase the risk of fatal injury for selected employees.

On May 25, 1983, a painter suffered fatal injuries when the suspended scaffolding from which he was working collapsed. The County Coroner requested NIOSH technical assistance to develop information on factors involved with the incident data.

CONTACTS/ACTIVITIES

After receiving notification, three Division of Safety Research personnel, a safety specialist, a safety engineer, and an epidemiologist, visited at the site to interview the employer and witnesses and to obtain comparison data from suitable co-workers. The research team, the police department, and the employer examined the impounded scaffold at an independent testing laboratory.

A debriefing session was held with the employer, other employees, and the contractor. During this introductory meeting, background information was obtained about the contractor and the employer, including an overview of their safety and health program. Interviews were conducted with witnesses and co-workers. Examining the scaffold assisted the researchers in developing hypotheses about the sequence of events leading to the incident.

SYNOPSIS OF EVENTS

The two workers had placed the scaffold supporting wire rope on the 7th floor permanently installed eye hooks. They then reeved the wire rope to the scaffold stirrups which are located at each end of the scaffold staging. After reeving was complete, the workers raised the scaffolding to the 7th floor windows. This action was accomplished by turning the drive motor directional switch to the "up" position and holding the motor switch in the "on" position.

The victim had to apply caulking around the windows. After caulking half way across the floor, he had to change positions, including independent life lines with a co-worker, who survived the incident. After caulking the remaining windows, the workers switched positions again in order to begin their descent.

The co-worker stated that he turned away from the victim and faced his stirrup in preparation of descent. As he did this, he felt some movement in the scaffold. He turned and looked at the victim, who motioned by hand signal to turn the directional switch to the "down" position. The co-worker signaled "okay" and turned to face his stirrup. As he was in the process of preparing his stirrup for downward movement plus getting his lanyard grab device ready to move down, he felt several sudden jerks and was suddenly dangling from his life line. After regaining his composure, the co-worker looked for the victim in the area of his life line. The co-worker then noticed the victim lying in the street across from the building.

GENERAL CONCLUSIONS AND RECOMMENDATIONS

There is some evidence which indicates the deceased was not familiar with the operation of this type of scaffold. For this type of scaffold, the operator must operate the drill and a brake lever at the same time with one hand, while releasing his lanyard on the safety line with the other hand.

Additionally, the victim's lanyard failed to prevent the fatal fall for one of two reasons. Either the lanyard was deteriorated to the extent that the impact load was in excess of the lanyard strength or the lanyard became entangled in the scaffold components.

It is suspected that the wire rope broke because the hoist's secondary safety mechanism did not function quickly enough. The wire rope broke at a level 20+ feet below where the scaffold was originally positioned. When the mechanism finally activated, the force of the falling scaffold caused the emergency braking cam to squeeze the rope to such an extent that it actually cut 5 of the 6 strands. The remaining strand was not of sufficient strength to hold the falling scaffold and it also broke.

It is recommended that workers who use scaffolds should be trained in the proper use, maintenance, and limitations of scaffolding, life lines and lanyards. Also management should be aware of their responsibilities when their workers are using scaffolds. Safety requirements for scaffolding are outlined in the OSHAct regulations 1910.28, 1910.29 and 1926.451.

Accidents Linked to Design

- 22% of 226 injuries that occurred from 2000 to 2002 in Oregon, Washington, and California were linked partly to design [Behm 2005]

- 42% of 224 fatalities in U.S. between 1990 and 2003 were linked to design [Behm 2005]

- In Europe, a 1991 study concluded that 60% of fatal accidents resulted in part from decisions made before site work began [European Foundation for the Improvement of Living and Working Conditions 1991]

- 63% of all fatalities and injuries could be attributed to design decisions or lack of planning [CHAIR safety in design tool 2001]

NOTES

Research conducted in the United States, Europe, and other regions has shown that design does affect the inherent risk in constructing a facility. Research linked design to 22% of injuries that occurred in western states and 42% of fatalities across the country. European researchers found that nearly two-thirds of fatalities and injuries were linked to design. Facility designers are encouraged to consult with occupational safety and health professionals early in the design process to identify and design out hazards and to reduce risk of injury, illness, and death.

SOURCES

Behm M [2005]. Linking construction fatalities to the design for construction safety concept. Safety Sci 43:589–611.

NOHSC [2001]. CHAIR safety in design tool. New South Wales, Australia: National Occupational Health & Safety Commission.

European Foundation for the Improvement of Living and Working Conditions [1991]. From drawing board to building site (EF/88/17/FR). Dublin: European Foundation for the Improvement of Living and Working Conditions.

Falls

Falls

- Number one cause of construction fatalities
  - in 2010, 35% of 751 deaths
    www.bls.gov/news.release/cfoi.t02.htm

- Common situations include making connections, walking on beams or near openings such as floors or windows

- Fall protection is required at height of 6 feet above a surface [29 CFR 1926.760].

- Common causes: slippery surfaces, unexpected vibrations, misalignment, and unexpected loads

NOTES

Falls are the number one cause of deaths in the construction industry. In 2004, 445 (36%) of 1,234 deaths were due to falls [BLS 2006]. By contrast, of 751 deaths in the construction sector in 2010, 35% were attributed to falls [BLS 2011a]. The decline in number of fatalities in the construction sector in 2010, compared to 2004, was attributed more to the economic downturn than to any other factor [BLS 2011b].

Falls from any height can be fatal. In construction, workers are often high off the ground. For structural reasons, the taller cross-sections of W shapes are usually chosen for beams. The flanges on W shapes may be less than six inches wide. Workers walk on beams, sometimes without fall protection. Fall protection is highly recommended and often required in most scenarios involving heights. OSHA requires fall protection at a height of 15 feet above a surface during steel erection. For other construction phases, it is 6 feet [29 CFR 1926.760].

SOURCES

BLS [2006]. Injuries, illnesses, and fatalities in construction, 2004. By Meyer SW, Pegula SM. Washington, DC: U.S. Department of Labor, Bureau of Labor Statistics, Office of Safety, Health, and Working Conditions [www.bls.gov/opub/cwc/sh20060519ar01p1.htm].

BLS [2011a]. Census of Fatal Occupational Injuries. Washington, DC: U.S. Department of Labor, Bureau of Labor Statistics. [www.bls.gov/news.release/cfoi.t02.htm].

BLS [2011b]. Injuries, Illnesses, and Fatalities (IIF). Washington, DC: U.S. Department of Labor, Bureau of Labor Statistics. [www.bls.gov/iif/home.htm]

OSHA [2001]. Standard number 1926.760: fall protection. Washington, DC: U.S. Department of Labor, Occupational Safety and Health Administration.

Death from Injury

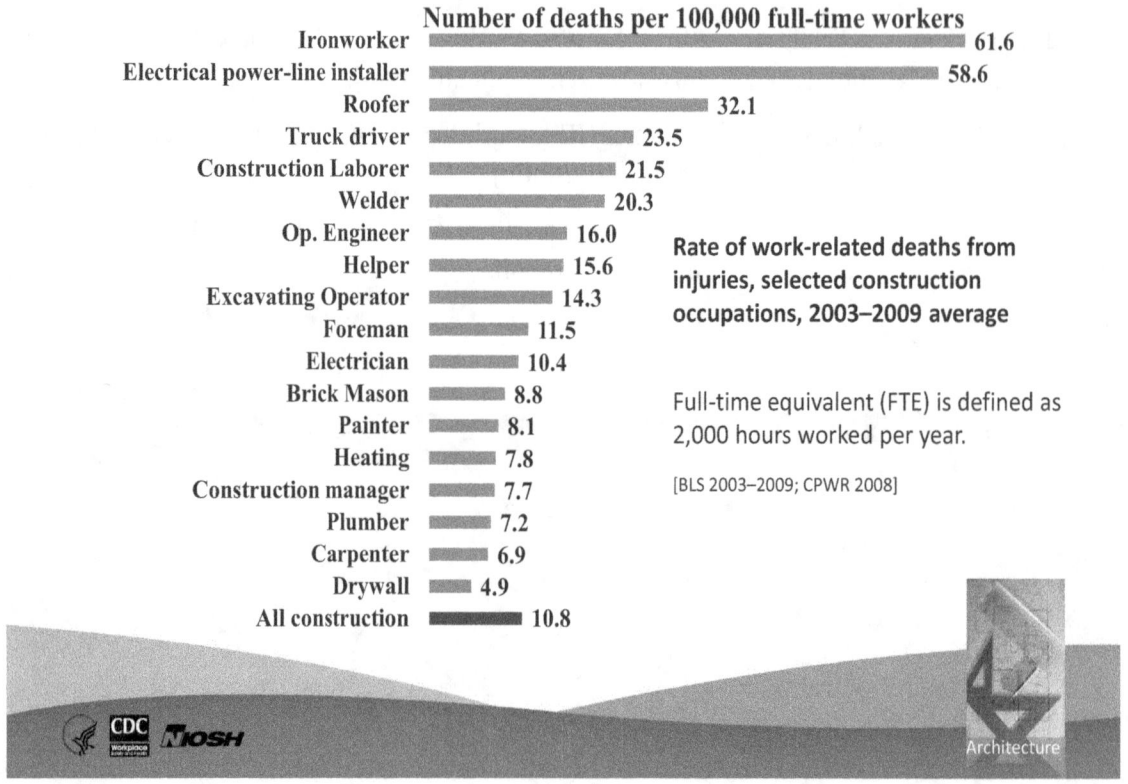

Number of deaths per 100,000 full-time workers

Ironworker	61.6
Electrical power-line installer	58.6
Roofer	32.1
Truck driver	23.5
Construction Laborer	21.5
Welder	20.3
Op. Engineer	16.0
Helper	15.6
Excavating Operator	14.3
Foreman	11.5
Electrician	10.4
Brick Mason	8.8
Painter	8.1
Heating	7.8
Construction manager	7.7
Plumber	7.2
Carpenter	6.9
Drywall	4.9
All construction	10.8

Rate of work-related deaths from injuries, selected construction occupations, 2003–2009 average

Full-time equivalent (FTE) is defined as 2,000 hours worked per year.

[BLS 2003–2009; CPWR 2008]

NOTES

The Center for Construction Research and Training compiles a "Construction Chart Book" using Bureau of Labor Statistics data [CPWR 2008]. It includes two illuminating charts useful for considering safety issues. This chart is compiled from 2003–2009 data on workplace fatalities. Ironworkers experience the highest work-related death rate, with 61.6 fatalities per 100,000 FTE.

SOURCES

BLS [2003–2009]. Census of Fatal Occupational Injuries. Washington, DC: U.S. Department of Labor, Bureau of Labor Statistics [www.bls.gov/iif/oshcfoi1.htm].

CPWR [2008]. The construction chart book. 4th ed. Silver Spring, MD: Center for Construction Research and Training.

Fatality Assessment and Control Evaluation

NIOSH FACE Program www.cdc.gov/niosh/face

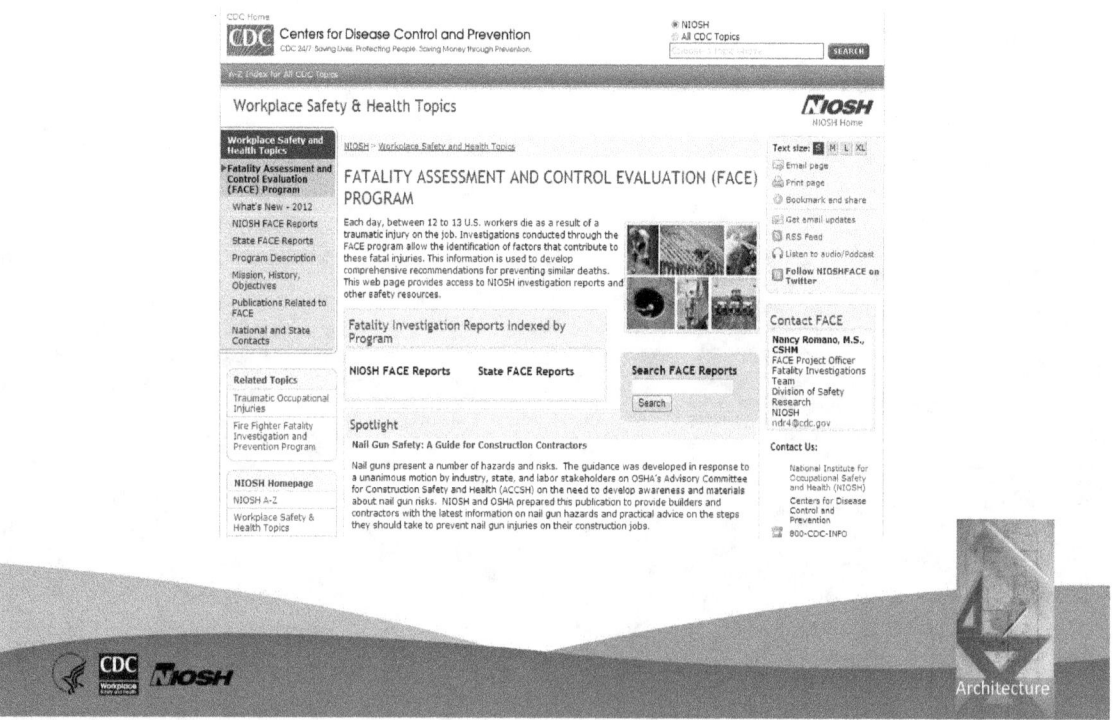

NOTES

The NIOSH Fatality Assessment and Control Evaluation Program examines worker fatalities by type of injury. By studying these reports, an enterprising designer can identify recurrent problems to "design out."

SOURCE

NIOSH Fatality Assessment and Control Evaluation Program [www.cdc.gov/niosh/face/]

What is Prevention through Design?

Eliminating or reducing work-related hazards and illness and minimizing risks associated with

- Construction

- Manufacturing

- Maintenance

- Use, reuse, and disposal of facilities, materials, and equipment

NOTES

PtD is a risk management technique that is being applied successfully in many industries, including manufacturing, healthcare, telecommunications, and construction. PtD is the optimal method of preventing occupational illnesses, injuries, and fatalities by designing out the hazards and risks. This approach involves the design of tools, equipment, systems, work processes, and facilities in order to reduce, or eliminate, hazards associated with work. The concept is simply that the safety and health of workers throughout the life cycle are considered while the product and/ or process is being designed. The life cycle starts with concept development, and includes design, construction or manufacturing, operations, maintenance, and eventual disposal of whatever is being designed, which could be a facility, a material, or a piece of equipment.

PtD processes have been required in other countries for several years now, but in the United States, PtD is being adopted on a voluntary basis. The National Institute for Occupational Safety and Health (NIOSH) is spearheading a national initiative in PtD and partnering with many professional organizations to apply the concept to their industry and professions. The Occupational Safety and Health Administration (OSHA) is very interested in PtD but is not currently considering making it mandatory.

PtD design professionals (that is, architects and/or engineers) working with the project owner (that is, the client) make deliberate design decisions that eliminate or reduce the risk of injuries or illness throughout the life of a project, beginning at the earliest stages of a project's life cycle. PtD is thus the deliberate consideration of construction and maintenance worker safety and health in the design phase of a construction project. PtD processes in construction have been required in the United Kingdom for over a decade and are being implemented in other countries such as Australia and Singapore.

PtD applies to the design of a facility, that is, to the aspects of the completed building that make a project inherently safer. PtD does not focus on how to make different methods of construction safer. For example, it does not focus on how to use fall protection systems, but it does include consideration of design decisions that influence how often fall protection will be needed. Similarly, PtD does not address how to erect safe scaffolding, but it does relate to design decisions that influence the location and type of scaffolding needed to accomplish the work. PtD concepts may also be used to design temporary structures. Some design decisions improve workplace safety. For example, when the height of parapet walls is designed to be 42", the parapet acts as a guardrail and enhances safety. When designed into the permanent structure of the building and sequenced early in construction, the parapet at this height acts to enhance safety during initial construction activities and during subsequent maintenance and construction activities, such as roof repair. In the United States, the employer is solely responsible for site safety.

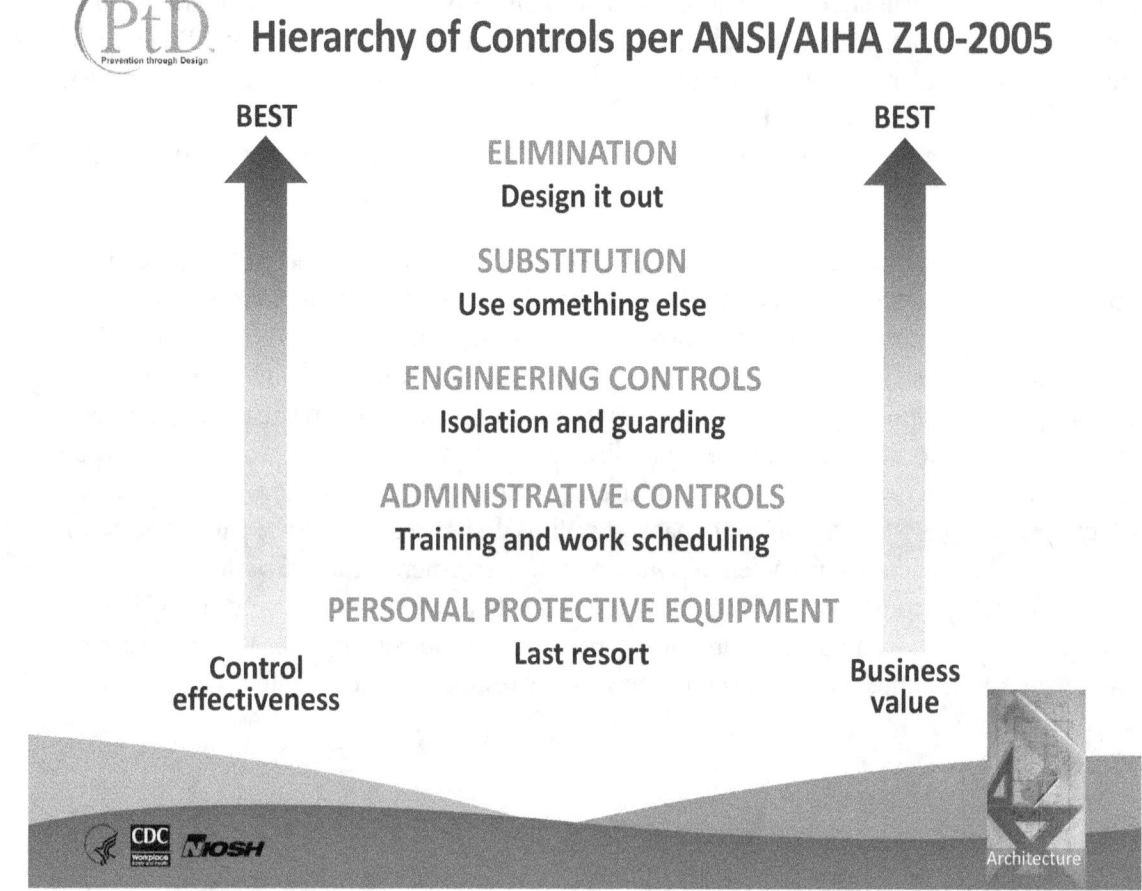

Hierarchy of Controls per ANSI/AIHA Z10-2005

BEST   BEST

ELIMINATION
Design it out

SUBSTITUTION
Use something else

ENGINEERING CONTROLS
Isolation and guarding

ADMINISTRATIVE CONTROLS
Training and work scheduling

PERSONAL PROTECTIVE EQUIPMENT
Last resort

Control effectiveness   **Business value**

NOTES

This slide shows the well-accepted Hierarchy of Controls. PtD anticipates and removes potential hazardous elements at the design phase of a project through elimination or substitution. Residual risks may be minimized through the use of engineering and administrative controls.

The top of the hierarchy is better in terms of improved occupational safety and health (OSH) and cost savings. Below is a description of the different levels, from most to least effective.

Elimination: "Design out" hazards and hazardous exposures.

Substitution: Substitute less-hazardous materials, processes, operations, or equipment. A larger crane may be specified when the load or the reach approaches the crane design limit. Nontoxic chemicals are preferred. The Green Chemistry movement replaces toxic compounds with less hazardous chemicals.

Engineering controls: Isolate process or equipment or contain the hazard. Remove hazard from work zone, e.g., with exhaust ventilation. Require two hands to operate machinery. Use warning devices to warn worker about entry into hazard zone. Signs, labels, alarms, and flashing lights give warnings. Safety switches, hand guards, and other engineering controls prevent certain kinds of injuries.

Administrative controls: Job rotation, work scheduling, training, well-designed work methods, and organization are examples. Administrative controls include training modules and company procedures. A well-organized worksite is safer than a messy one. Reducing the clutter on a construction site improves worker safety by reducing the exposure to hazards. The foreman controls site layout and housekeeping policies.

Personal Protective Equipment (PPE): Includes but is not limited to safety glasses for eye protection; ear plugs for hearing protection; clothing such as safety shoes, gloves, and overalls; face shields for welders; fall harnesses; and respirators to prevent inhalation of hazardous substances.

SOURCE

ANSI/AIHA [2005]. American national standard for occupational health and safety management systems. New York: American National Standards Institute, Inc. ANSI/AIHA Z10-2005.

Personal Protective Equipment (PPE)

- Last line of defense against injury
- Examples:
    - Hard hats
    - Steel-toed boots
    - Safety glasses
    - Gloves
    - Harnesses

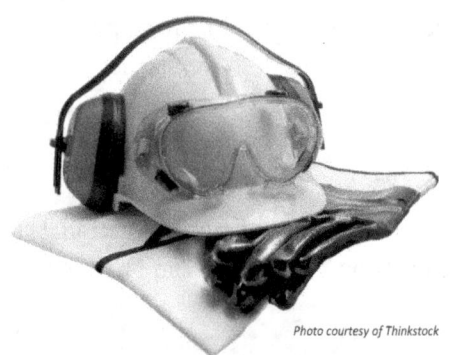

Photo courtesy of Thinkstock

OSHA www.osha.gov/Publications/osha3151.html

NOTES

Personal Protective Equipment, or PPE, includes items worn as a last line of defense against injury. OSHA-required PPE can include hardhats, steel-toed boots, safety glasses or safety goggles, gloves, earmuffs, full body suits, respiratory aids, face shields, and fall harnesses.

SOURCES

NOHSC [2001]. CHAIR safety in design tool. New South Wales, Australia: National Occupational Health & Safety Commission.

OSHA PPE publications

www.osha.gov/Publications/osha3151.html
www.osha.gov/OshDoc/data_General_Facts/ppe-factsheet.pdf
www.osha.gov/OshDoc/data_Hurricane_Facts/construction_ppe.pdf

Photo courtesy of Thinkstock

 **PtD Process**

[Hecker et al. 2005]

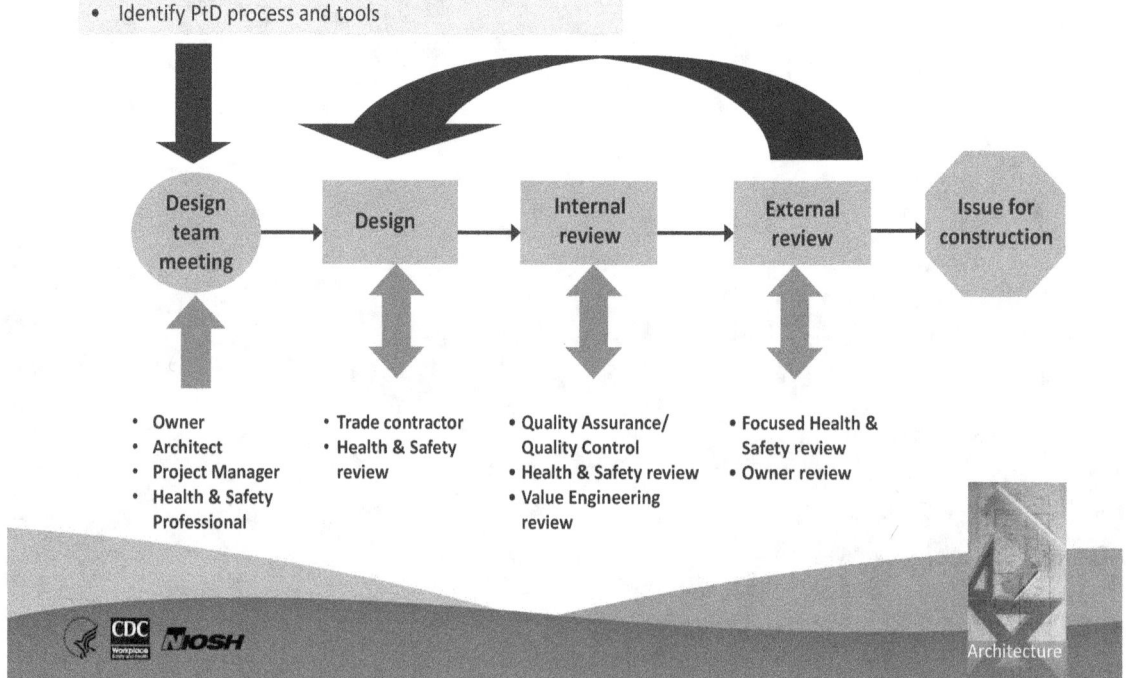

NOTES

This graphic depicts the typical PtD process. The key component of this process is the incorporation of safety knowledge into design decisions. For example, site safety should be considered throughout the design process. A progress review specifically focused on site safety may be effective. Site safety knowledge can be provided by trade contractors, an on-site employee, or a hired consultant. The graphic emphasizes the importance of communication between designers and constructors. Such communication during design may reveal steps to reduce construction duration.

Many project managers schedule a Value Engineering review prior to issuing drawings for bid. The purpose is to reduce overall project costs. Unfortunately, during the review, redundant systems that are necessary to protect worker health may be eliminated. It is therefore considered a best practice to conduct a focused Health & Safety (H&S) review before drawings are issued.

SOURCE

Hecker S, Gambatese J, Weinstein M [2005]. Designing for worker safety: moving the construction safety process upstream. Prof Saf *50*(9):32–44.

Integrating Occupational Safety and Health with the Design Process

Stage	Activities
Conceptual design	Establish occupational safety and health goals, identify occupational hazards
Preliminary design	Eliminate hazards, if possible; substitute less hazardous agents/processes; establish risk minimization targets for remaining hazards; assess risk; and develop risk control alternatives. Write contract specifications.
Detailed design	Select controls; conduct process hazard reviews
Procurement	Develop equipment specifications and include in procurements; develop "checks and tests" for factory acceptance testing and commissioning
Construction	Ensure construction site safety and contractor safety
Commissioning	Conduct "checks and tests," including factory acceptance; pre–start up safety reviews; development of standard operating procedures (SOPs); risk/exposure assessment; and management of residual risks
Start up and occupancy	Educate; manage changes; modify SOPs

NOTES

The integration of OSH goals within the design processes is an essential concept because it elevates the importance of safety and health as a value proposition in the overall design, construction, and operation of projects.

Identify hazards during conceptual design. Follow the Hierarchy of Controls to eliminate or reduce risks.

For example, how much space is needed to access, maintain, and replace HVAC units?

Use project specifications to require the inclusion of fall protection systems such as permanent anchor points for lifelines. Reduce fall hazards by specifying a ladder-free construction site.

Obtain a site plan that shows the location of existing underground and overhead utilities and develop traffic control plans to avoid those hazards.

Compare the list of desirable safety features against the detailed design.

Obtain feedback from safety and health professionals, contractors, and trade representatives. Modify the design to improve safety.

Call out required hazard controls on the drawing and in the contract specifications when possible. During procurement, compare materials and equipment received against the contract specifications. Develop a checklist for commissioning.

During construction, how do contractors communicate with the project manager and each other? Who has the authority to correct a hazardous condition on the worksite?

What procedures are followed before and after permanent equipment reaches the site? Follow the commissioning checklist!

Does the building have unusual features? Educate the owners and tenants.

Are special operating procedures required?

At each stage of the design process, think of ways to reduce the workplace risks.

Safety Payoff During Design

[Adapted from Szymberski 1997]

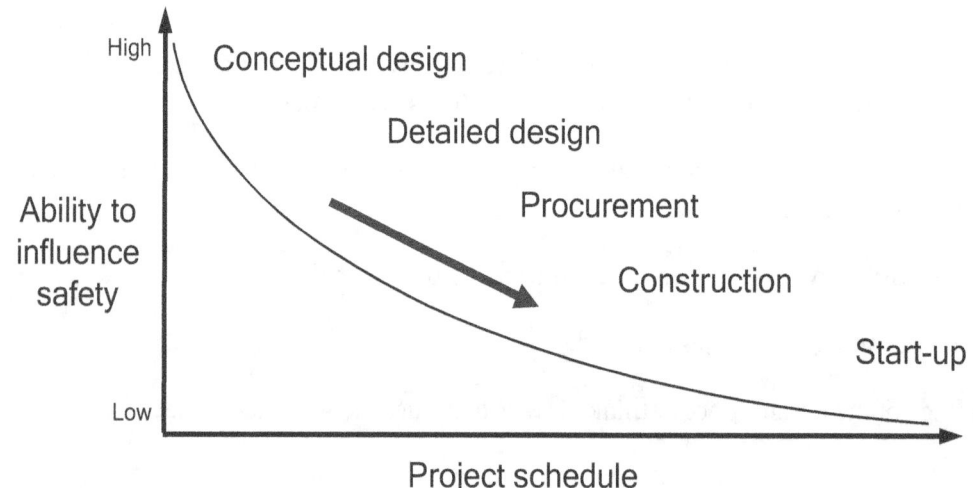

NOTES

Most owners and design professionals know intuitively that the earlier in the design process that cost is considered, the easier it is to achieve cost-effective goals. The same is true for construction duration and quality. A worker's ability to influence project criteria decreases as the design and construction progress. The same principle is true for construction safety. The earlier in the project life cycle that safety is considered, the easier it is to reduce hazards. This concept is in contrast to the prevailing methods of planning for construction site safety, which do not begin until a short time before the construction phase, when the ability to influence safety is limited.

SOURCE

Szymberski R [1997]. Construction project planning. TAPPI J *80*(11):69–74.

PtD Process Tasks

[Adapted from Toole 2005; Hinze and Wiegand 1992]

- Perform a hazard analysis

- Incorporate safety into the design documents

- Make a CAD model for member labeling and erection sequencing

Photo courtesy of Thinkstock

NOTES

This slide provides more details about the PtD process. Before, during, or after the conceptual design of a building, a hazard analysis can be performed. The designer meets with field professionals to review constructability, looking through the entire design for any hazards and addressing those hazards. The field professional can teach an inexperienced designer how to minimize risks in the field.

The safety input received during conceptual design can be reflected in detailed design drawings and specifications. Another constructability review should occur as the detailed design nears completion.

Sometimes the drawings that result from a PtD process look the same as typical construction drawings, but they are inherently safer for construction. Other times, drawings include special details and labels to make it easier for workers to erect the design safely.

Construction documents can be supplemented with graphic models and tables that contribute to safe erection. For example, a CAD file can be used to label steel members for safe erection sequencing. New software such as building information modeling (BIM) is able to show the final layouts of buildings and can detect any spatial problems before construction starts. Clearly labeled shop drawings eliminate confusion during installation. The BIM program can recommend efficient, safer erection sequencing.

SOURCES

Hinze J, Wiegand F [1992]. Role of designers in construction worker safety. Journal of Construction Engineering and Management *118*(4):677–684.

Toole TM [2005]. Increasing engineers' role in construction safety: opportunities and barriers. Journal of Professional Issues in Engineering Education and Practice *131*(3):199–207.

Photo courtesy of Thinkstock

Designer Tools

- Checklists for construction safety [Main and Ward 1992]

- Design for construction safety toolbox [Gambatese et al. 1997]

- Construction safety tools from the UK or Australia
  - Construction Hazard Assessment Implication Review (CHAIR) [NOHSC 2001]

NOTES

Most designers are not trained in PtD or construction site safety. It is therefore critical that they be given tools to facilitate the process. A PtD checklist alerts designers to common design elements that can lead to unnecessary hazards and identifies design options that are inherently safer. An example checklist is provided on the next slide.

The Design for Construction Safety Toolbox was developed by a Construction Industry Institute–sponsored research team that included leading PtD academics. This toolbox was recently updated by Professor Jimmie Hinze at the University of Florida. The United Kingdom and Australia make available on the Web valuable PtD tools that reflect their experiences with PtD legislation and voluntary initiatives. For example, CHAIR (Construction Hazard Assessment Implication Review) is an Australian tool and methodology that systematically combines brainstorming and decisions to gradually rid the design of unnecessary hazards.

SOURCES

NOHSC [2001]. CHAIR safety in design tool. New South Wales, Australia: National Occupational Health & Safety Commission.

Gambatese JA, Hinze J, Haas CT [1997]. Tool to design for construction worker safety. J Arch Eng 3(1):2–41.

Main BW, Ward AC [1992]. What do engineers really know and do about safety? Implications for education, training, and practice. Mechanical Engineering 114(8):44–51.

Example Checklist

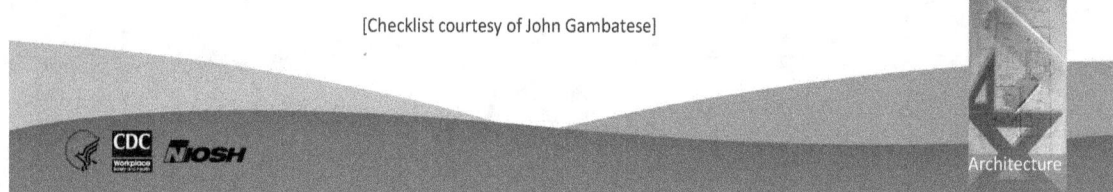

Item	Description
1.0	**Structural Framing**
1.1	Space slab and mat foundation top reinforcing steel at no more than 6 inches on center each way to provide a safe walking surface.
1.2	Design floor perimeter beams and beams above floor openings to support lanyards.
1.3	Design steel columns with holes at 21 and 42 inches above the floor level to support guardrail cables.
2.0	**Accessibility**
2.1	Provide adequate access to all valves and controls.
2.2	Orient equipment and controls so that they do not obstruct walkways and work areas.
2.3	Locate shutoff valves and switches in sight of the equipment which they control.
2.4	Provide adequate head room for access to equipment, electrical panels, and storage areas.
2.5	Design welded connections such that the weld locations can be safely accessed.

[Checklist courtesy of John Gambatese]

NOTES

Like many PtD checklists, this example includes hazards associated with both construction and maintenance.

SOURCE

Checklist courtesy of John Gambatese

 **Why Prevention through Design?**

- Ethical reasons

- Construction dangers

- Design-related safety issues

- Financial and non-financial benefits

- Practical benefits

Photo courtesy of Thinkstock

NOTES

Engineers have strong ethical reasons to apply the PtD concept to their designs. There are practical benefits, too. Lost-time accidents delay the job, destroy crew morale, and cost money. The next few slides will show there are many reasons why owners and design professionals should be motivated to incorporate PtD in a project.

SOURCE

Photo courtesy of Thinkstock

Ethical Reasons for PtD

- National Society of Professional Engineers' Code of Ethics:

    "Engineers shall hold paramount the safety, health, and welfare of the public..."

- American Society of Civil Engineers' Code of Ethics:

    "Engineers shall recognize that the lives, safety, health and welfare of the general public are dependent upon engineering decisions..."

NSPE www.nspe.org/ethics/index.html

ASCE www.asce.org/content.aspx?id=7231

NOTES

Many safety professionals and design professionals believe that PtD is clearly an ethical duty. Nearly all national engineering societies include in their code of ethics a statement similar to the one shown here for the National Society of Professional Engineers: "Engineers shall hold paramount the safety, health, and welfare of the public."

The American Society of Civil Engineers goes one step further and explicitly states that engineering decisions directly affect safety. These organizations pledge to protect "the public." Why? The public lacks the knowledge of forces, stresses, and other risk-related issues that contribute to hazardous work-related conditions. Many construction and maintenance workers, especially apprentices, fail to perceive an unsafe condition. Even if construction workers recognize a hazard that could have been eliminated or reduced through an alternative design, there are significant barriers to redesign after construction is under way. Their safety and health deserve consideration.

SOURCES

American Society of Civil Engineers [ASCE] [www.asce.org/Content.aspx?id=7231]

National Society of Professional Engineers [NSPE][www.nspe.org/ethics]

PtD Applies to Constructability

- How reasonable is the design?
  - Cost
  - Duration
  - Quality
  - Safety

Photo courtesy of the Cincinnati Museum Center www.cincymuseum.org

NOTES

Most designers know that what may look great on paper might not be constructible. An important part of the design process is to evaluate the design's constructability, that is, to what extent the design can be constructed at a reasonable price, quickly, and with high quality. Safety is an important part of constructability. Accidents cost money, delay construction, and may result in bad publicity rather than acclaim for the owner.

Exciting buildings designed by creative architects require strong consideration of worker safety and health early in the design process. Owners realize these one-of-a-kind structures cost more to build and generally present unique challenges for the construction crew. Fewer construction firms have the expertise needed to build the structure, so fewer firms submit a bid, which reduces competition and therefore drives up price, resulting in higher bond and insurance costs. The timeline for procurement and construction is harder to estimate. The uniqueness of the design creates construction and maintenance challenges. Unusual materials, custom fabrications, non-standard specifications, and striking aesthetic features inherent in these designs require greater collaboration. The PtD process shown on the next slide helps the design team identify potential hazards in time to devise appropriate prevention strategies for construction crews and future maintenance workers. The project manager should include occupational safety and health professionals throughout the design process to design-in protections for workers.

SOURCE

Photo courtesy of the Cincinnati Museum Center

 **Business Value of PtD**

- Anticipate worker exposures—be proactive

- Align health and safety goals with business goals

- Modify designs to reduce/eliminate workplace hazards in

Facilities	Equipment
Tools	Processes
Products	Work flows

Improve business profitability!

AIHA www.ihvalue.org

NOTES

Companies that have implemented PtD programs experience lower than average injury and illness rates and lower workers' compensation expenses. However, the business value of PtD does not end there. In a study entitled Demonstrating the Business Value of Industrial Hygiene (known as The Value Study), findings showed that significant business cost savings accrue when hazards are eliminated or reduced.

SOURCE

American Institute of Industrial Hygienists [AIHA] [2008]. Strategy to demonstrate the value of industrial hygiene [www.aiha.org/votp_NEW/pdf/votp_exec_summary.pdf].

 **Benefits of PtD**

- Reduced site hazards and thus fewer injuries

- Reduced workers' compensation insurance costs

- Increased productivity

- Fewer delays due to accidents

- Increased designer-constructor collaboration

- Reduced absenteeism

- Improved morale

- Reduced employee turnover

NOTES

PtD yields better value for owners and better health for the workers. When a project is designed with construction worker safety in mind, there are fewer hazards on site, with fewer injuries and fatalities. A reduction in injuries results in reduced workers' compensation insurance and less down-time, a direct savings for the employer. Experience shows PtD increases productivity and reduces labor costs. Safer designs lead to fewer project delays.

 **Industries Use PtD Successfully**

- Construction companies
- Computer and communications corporations
- Design-build contractors
- Electrical power providers
- Engineering consulting firms
- Oil and gas industries
- Water utilities

  And many others

NOTES

Major corporations in diverse industries and public utilities in several states have applied PtD through initiatives or established programs. At these companies, worker safety and health are an integral part of the corporate culture. International construction firms first encountered PtD on their European projects. They brought the concepts and related cost savings home to their American operations. Many firms provide PtD training for their design engineers in the areas of construction site safety, PtD checklists, and safety constructability reviews. These firms want to hire engineers who have a basic understanding of PtD.

Site Planning

ARCHITECTURAL DESIGN AND CONSTRUCTION

Site Planning

NOTES

Depending on the project delivery method, the site utilization/layout planning may or may not involve collaboration of the designer and the contractor. Designers will probably be involved in defining the construction entrance and staging area, and they may need to work closely with the constructor to develop the site utilization plans and layout logic.

Site planning hazards to consider during the design phase include

- overhead and underground utilities
- construction and other vehicle movement in relationship to the specified work
- health hazards such as dust emissions from nearby work or from contaminated land
- lifts above local areas
- general worker and material access issues.

In this section, we will cover some examples of hazards that can be identified in the design phase and ways to mitigate or reduce the hazards to workers.

Site Location and Access

- Materials
- Workers
- Equipment
- Pedestrians

Photo courtesy of Thinkstock

NOTES

Proper planning in the design phase includes considering transportation of materials and workers as well as vehicle traffic. The Traffic Control Plan (TCP) separates through traffic from construction traffic. On highway sites, half of the deaths are due to construction equipment, specifically backovers and rollovers.

Lifting equipment provides a great benefit to the health and safety of construction workers; it reduces physical lifting and eases access to materials. However, the use of large cranes requires site planning in relation to roads, neighborhoods, and other businesses. Their positioning on the site will affect the safety of workers as well as the general public.

Positioning equipment at the site should involve these design considerations:

1. Consider the existing site and its potential hazards in relation to the heavy equipment required to perform the scope of work. Warn and inform constructors about the potential hazards. Consider moving the work to other locations within the site.

2. Review the site and the specified work to identify potential risks that are unacceptable.

SOURCE

Photos courtesy of Thinkstock

Prefabrication

- Prefabrication and preassembly will likely increase worker safety [Haas 2000]

- Prefabrication reduces work at height [CIRIA 2004]

- Prefabrication may reduce cold/heat stress

- Prefabrication increases heavy lifting; possible access and transportation issues

    – Managing risks is the key

NOTES

Prefabrication usually is considered on the basis of a variety of factors, including duration, quality, cost, site storage, and access conditions. Various design considerations include the size and complexity of prefabricated elements, unloading, and installation. For example, the use of panelized walls in U.S. homebuilding involves unloading, organizing, moving, and installing the panels, which may involve varying degrees of ergonomic hazard. Failing to prequalify or evaluate the expertise of off-site prefabrication subcontractors can result in shifting worker injuries onto the subcontractors' log, rather than actually preventing injuries. Large prefabricated components can introduce significant transportation risks. The use of larger prefabricated metal panels, such as in commercial buildings, may involve lifting hazards and the need for coordination with other trades. Prefabrication is safer, faster, and cheaper than building those elements on site.

The UK's Construction Industry Research and Information Association (CIRIA) gives this example of how prefabrication helps safety. "Designers can reduce the need to work at height when erecting a steel frame by designing the steel work in modular sections, which can be prefabricated at ground level and sequentially lifted into place. This does not eliminate working at height entirely but should reduce it significantly. The safety of those who have to work at height can further be enhanced if the ground slab is installed before the steel frame to provide a stable, flat working surface from which mobile elevating work platforms can be used for bolting up. This is much safer than using ladder access and quicker than scaffolding. Although designers are not expected to specify particular construction methods or sequences, they will be expected {according to the UK's CDM} to have considered possible alternatives when the hazards are being identified. If the assumptions of construction methods and sequence become inextricably woven into the design, such that there is only one reasonable choice, then this will have to be made known" [CIRIA 2004].

Design students should have a global perspective. In the United States, opportunities within design-build firms may expose designers to the philosophy described by CIRIA. The next few slides show deliveries of prefabricated elements and highlight some of the hazards.

SOURCES

Construction Industry Research and Information Association (CIRIA) [2004]. CDM regulations: work sector guidance for designers. 2nd Ed. London: CIRIA.

Haas C, O'Connor J, Tucker R, Eickmann J, Fagerlund W [2000]. Prefabrication and preassembly trends and effects on the construction workforce. Report no. 14. Austin, TX: Center for Construction Industry Studies, The University of Texas at Austin.

Site Activities Case Study

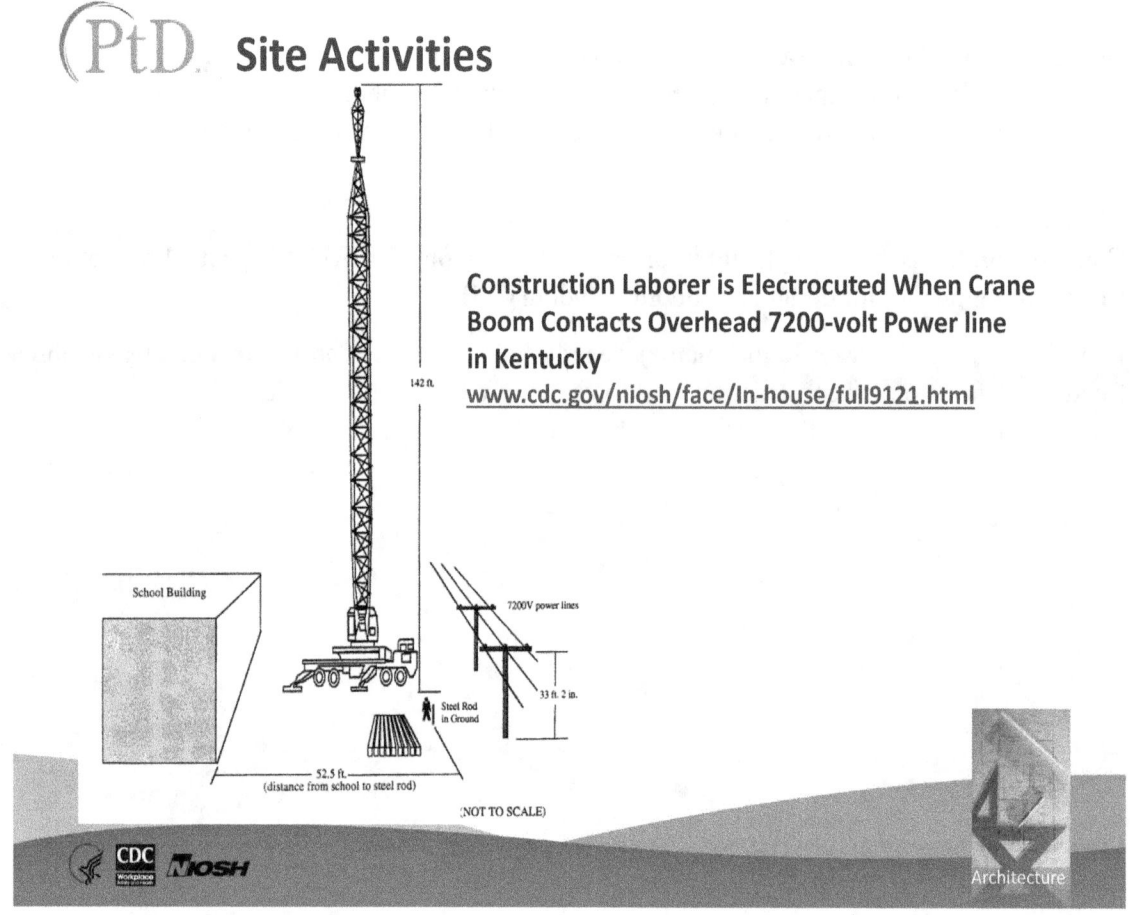

NOTES

This FACE report describes an incident resulting in the death of a laborer who was holding a tag line attached to a load when the crane boom touched a power line.

SOURCES

NIOSH Fatality Assessment and Control Evaluation (FACE) Program [1991]. Construction laborer is electrocuted when crane boom contacts overhead 7200-volt power line in Kentucky. FACE 9121 [www.cdc.gov/niosh/face/In-house/full9121.html].

FACE report courtesy of the Commonwealth of Kentucky Division of Safety Research

Construction Laborer is Electrocuted When Crane Boom Contacts Overhead 7200-volt Powerline in Kentucky

SUMMARY

A 37-year-old construction laborer (victim) was electrocuted while pulling a wire rope load choker attached to a crane cable toward a load. The choker was to be connected to a steel roof joist which was to be lifted 150 feet across the roof of a one-story school and set in place. The cab of the crane was positioned 11 feet 6 inches from a three-phase 7200-volt powerline. After a previous roof joist had been moved, the crane operator swung the crane boom and cable back toward the victim. The victim grasped the choker in his left hand and with his right hand held onto a steel rod that had been driven into the ground nearby. At this point, the crane cable contacted the powerline and the electrical current passed across the victim's chest and through the steel rod to ground, causing his electrocution. NIOSH investigators concluded that, in order to prevent future similar occurrences, employers should:

- *comply with existing regulations pertaining to clearance distances between cranes and powerlines*

- *use a designated signal person when necessary*

- *evaluate a jobsite prior to the start of work to determine the safest areas for material storage, machinery placement during operations, and the size and type of machinery to be used*

- *contact the local electric utility when work is to be performed in proximity to overhead powerlines*

- *instruct employees to use non-conductive links, chokers, or taglines when working in proximity to overhead powerlines.*

INTRODUCTION

On June 24, 1991, a 37-year-old construction laborer was electrocuted when the crane cable connected to the wire rope choker he was holding contacted a 7200-volt powerline. On July 25, 1991, officials of the Kentucky Occupational Safety and Health Administration notified the Division of Safety Research (DSR) of the death, and requested technical assistance. On July 31, 1991, a DSR safety specialist, safety engineer, and medical officer traveled to the incident site to conduct an investigation. The incident was reviewed with OSHA officials, employer representatives, and the county coroner's office. The autopsy report, medical records, and photographs of the incident site were obtained.

The employer in this incident was a crane rental service that also provided services in steel erection and demolition. The employer had been in business for 6 years and employed 15 workers. The company owner managed the safety function as a collateral duty. Meetings were conducted prior to the start of each project to discuss the safety considerations associated with that project. Additionally, monthly safety meetings for all employees were held at the company office and tailgate safety meetings were conducted at jobsites. Training was accomplished on-the-job. Company workers were aware of OSHA regulations regarding the clearance between cranes and powerlines.

INVESTIGATION

The company had been subcontracted to install steel roof joists and roof decking above the existing roof of a one-story school building. Steel columns to support the joists had been installed through the roof by another contractor. The new roof would raise the height of the one-story structure by 4 feet. The prime contractor's 50-ton conventional crane with a 190-foot-long boom and jib was used to lift the joists and set them in place. The company had a 50-ton hydraulic crane with a 150-foot-long boom at the site. The owner felt that the conventional crane, because of its greater lifting capacity, would be the safer machine to use for this particular job. The crane was positioned between the school and a three-phase, 7200-volt powerline--11 feet from the powerline (Figure). The distance between the school and the powerline was 58.5 feet. Two stacks of joists had been placed between the powerline and school, one 14 feet from the powerline, and the other 32 feet from the first stack and 12 feet from the school. The lengths of joists ran parallel to the powerline.

The day before the incident, the crew had begun to set the joists on the far side of the roof, approximately 150 feet away from the crane. The crew consisted of a crane operator, three laborers on the roof setting the joists, and one laborer on the ground (victim) connecting the joists to the crane. The crew set the joists the entire day without incident.

On the day of the incident, the crew began setting the joists on the side of the roof away from the crane. The crane operator lifted a joist from the stack nearest the school, swung it across the roof, and began to set it in place when the laborers noticed that it was the wrong length. The operator returned the joist to the stack nearest the school, where the victim unhooked it. The operator then swung the boom toward the stack of joists nearest the powerline. The victim grabbed the choker (a short length of wire rope with eyes spliced into either end; it was designed to be wrapped around a load, threaded through itself, and hooked to a crane hook) with his left hand and began to pull the choker and crane cable toward the stack of joists and away from the powerline. As the victim grabbed a steel rod that had been driven into the ground with his right hand (possibly to steady himself), the crane cable contacted the powerline 36 feet above the end of the choker. The electrical current passed down the cable, across the victim's chest, and down the steel rod to ground, causing the victim's electrocution.

A worker on the roof was certified in cardiopulmonary resuscitation (CPR) and initiated CPR within a minute. The emergency medical service was summoned, and transported the victim to the hospital, where he was pronounced dead on arrival. The body displayed burn marks consistent with death by electrocution.

During interviews immediately following the incident, the crane operator stated that he did not know how close the boom of the crane was to the powerline, since he was watching the ball at the end of the crane cable. The operator was maneuvering the ball so as not to hit the victim, who was 97 feet from the body of the crane. The cable's length from the end of the boom to the cable hook was 142 feet. It is assumed that counter forces on the cable--the boom swinging the cable in one direction and the victim pulling the cable in the opposite direction--caused the cable to whip into the powerline. Although the victim was standing 10 feet from the power pole, a scale drawing of the area demonstrates that with the crane positioned 11 feet from the powerline and its 190-foot boom positioned at a 70-degree angle with 142 feet of cable extended, the ball would be 10 feet from the power pole but only 5 to 7 feet from the powerline at 33 feet above ground level--the height of powerline.

CAUSE OF DEATH

The medical examiner ruled the cause of death as accidental electrocution with cardiorespiratory arrest.

RECOMMENDATIONS/DISCUSSION

Recommendation #1: Employers should ensure that employees comply with existing regulations pertaining to clearance distances between cranes and powerlines.

Discussion: OSHA regulations 29 CFR 1926.550 (a)(15) and 1910.180 (j) require that the minimum clearance between electric lines rated 50 kV or below and any part of the crane or load shall be 10 feet, unless the electrical lines have been "de-energized and visibly grounded" at the point of work or physical contact between the lines, equipment, or machines is "prevented by the erection of insulating barriers which cannot be part of the crane."

Recommendation #2: Employers should designate a worker as a signal person if it is difficult for the crane operator to maintain clearance by visible means.

Discussion: OSHA regulation 29 CFR 1926.550 (a) (15) (IV) requires that a person be designated to observe clearance of the equipment and to give timely warning for "all" operations where it is difficult for the operator to maintain desired clearances by visual means. In this instance, the operator's attention was focused on the ball on the end of the crane and the victim 97 feet away, not the clearance between the crane boom and the powerline.

Recommendation #3: Employers should evaluate a jobsite prior to the start of any project involving the use of construction machinery, such as a crane, to identify the safest areas for the storage of materials, the placement of machinery during operations, and the type and size of machinery to be used.

Discussion: During the planning stages of a project, a comprehensive workplace assessment should be conducted by qualified professionals to identify the appropriate size and type of machinery, safest areas for material storage, and the proper position for machinery during operations. If the areas had been identified during the planning phases, it may have been possible to stack the steel roof joists on the opposite side of the school where the powerline hazard could have been eliminated. The joists could still have been lifted 150 feet across the top of the roof to the far side, but the powerline would have been 52 feet away. It might also have been possible to use a smaller crane on each side of the school to position the steel joists. The figure of the crane drawn to scale demonstrates that with the 190-foot-long boom of the crane at a 70-degree angle and 142 feet of cable extended, the ball and choker are 10 feet from the power pole at ground level but only 5 to 7 feet from the powerline at 33 feet above ground (height of powerline). In this instance, the crew may have believed that a 10-foot clearance was maintained. By evaluating the distance and height of the lift, given the crane boom angle and height, potential hazards associated with overhead powerlines can be identified and controlled.

Recommendation #4: Employers should contact the local electric utility when work is to be performed in proximity to overhead powerlines.

Discussion: When work is to be performed in close proximity to overhead powerlines, employers should contact the local electric utility to discuss the work that is to be performed and what safety measures, if any, need to be enacted. In this instance, covering the phase of the powerline nearest the crane with insulated line hoses would have reduced the severity of, and the exposure to, the electrical hazard.

Recommendation #5: Employers should instruct workers to use nonconductive links, chokers, and/or taglines when guiding or hooking loads near overhead powerlines.

Discussion: When cranes are scheduled for use in work areas where overhead powerlines are present, employers should consider installing nonconductive links between the lifting cable and the breaker ball/hook assembly. Nonconductive chokers wrapped around loads and connected to the hook assembly provide an additional measure of worker protection. Employers also should instruct workers that nonconductive taglines should be used when hooking or guiding loads near overhead powerlines. Dry polypropylene rope is an excellent material for use as a nonconductive tagline.

REFERENCES
29 CFR 1926.550 (a)(15) Code of Federal Regulations, Washington, DC: U.S. Government Printing Office, Office of the Federal Register.
29 CFR 1910.180 (j) Code of Federal Regulations, Washington, DC: U.S. Government Printing Office, Office of the Federal Register.

29 CFR 1926.550 (a)(15)(IV) Code Of Federal Regulations, Washington, DC: U.S. Government Printing Office, Office of the Federal Register.

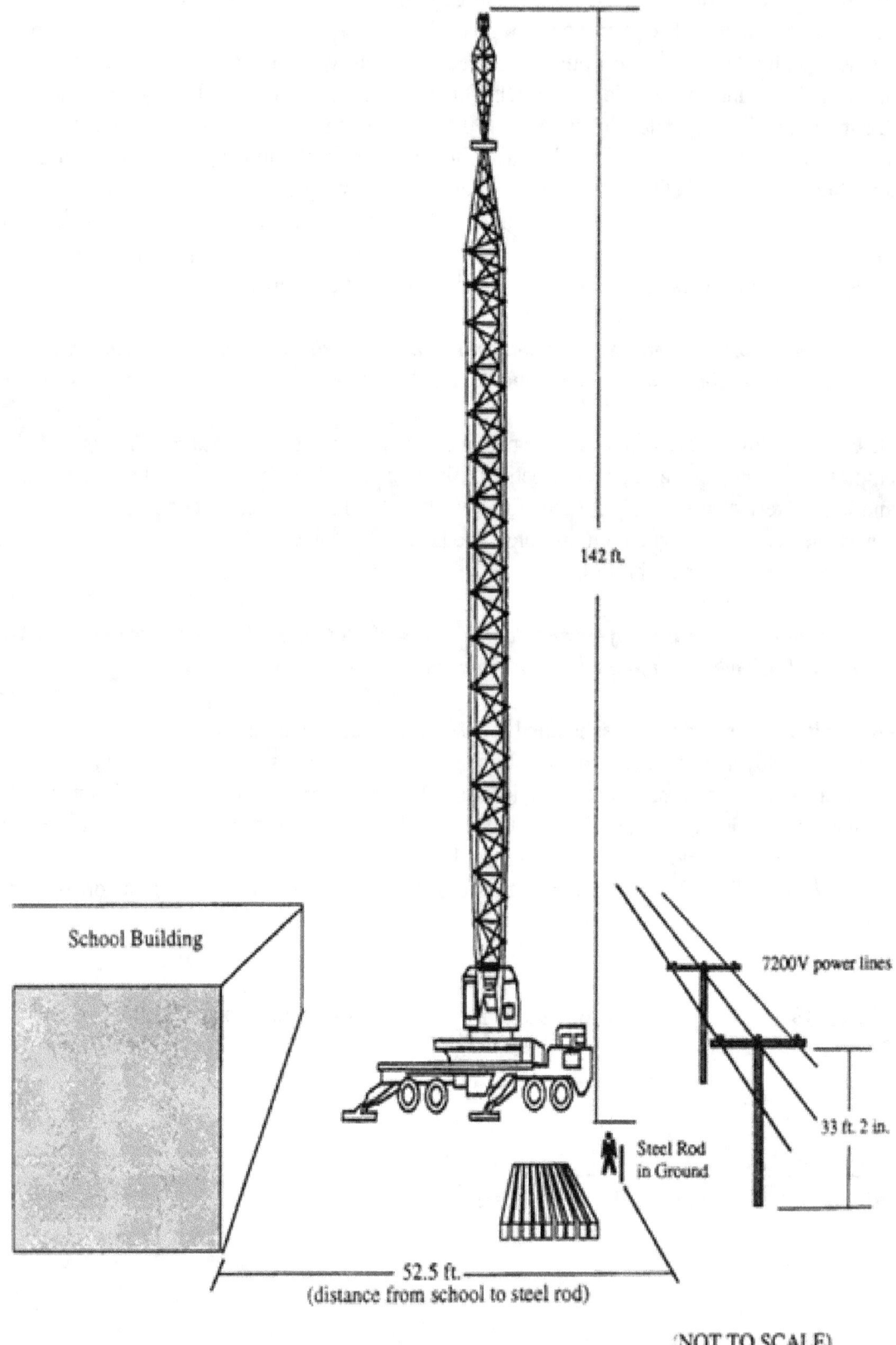

School Building

142 ft.

7200V power lines

Steel Rod
in Ground

33 ft. 2 in.

52.5 ft.
(distance from school to steel rod)

(NOT TO SCALE)

Figure *Configuration of the Incident Site*

Cranes and Derricks

- Carefully plan erection and disassembly

- Site layout affects crane maneuverability

- Show site utilities on plans

- Comply with OSHA standards

Photo courtesy of Walter Heckel

OSHA comprehensive crane standard: www.osha.gov/FedReg_osha_pdf/FED20100809.pdf
Regulation text: www.osha.gov/cranes-derricks/index.html

NOTES

Among other tasks, cranes are used to lift structural members and equipment into place. Cranes are the most complex machines on a construction site. Crane erection and disassembly must be carefully planned.

Where do you place the crane? Ideally, the crane can lift all members from one location without interfering with any other operations. The biggest danger in site layout is overhead power lines. Site contingency plans should include telephone numbers to contact the power company, as well as numbers to summon emergency personnel. Although it is the contractor's responsibility to deal with power lines, the designer can help by including the power line locations on the plans.

Overturning is often the result of moments created by the load. Cherry pickers are particularly susceptible. Cranes operate within a range defined by the mass of the crane, the length of the boom, and the mass of the load. Operators may be tempted to extend the boom a few more feet to pick up a load, when it would be safer to move the crane closer. As the load is lifted, the crane tips. Critical lifts near the crane load capacity or lifts involving two crane picks should be carefully planned.

Another hazard is boom collapse. In this instance, the lift exceeds the design limits of the boom. There is always the possibility that the operator will lose control of the load, especially when it is windy. A swinging load may impact adjacent structures or touch a power line. In several instances, the crane operator died when the load swung back into the cab.

The next two slides highlight two best practices for using a crane to load/unload trucks.

SOURCES

The OSHA comprehensive crane standard is available at www.osha.gov/FedReg_osha_pdf/FED20100809.pdf.

The regulation text is available at www.osha.gov/cranes-derricks/index.html.

A press release for the standard can be found at www.advancedsafetyhealth.com/blog/index.php/category/cranes.

Photo courtesy of Walter Heckel

Center the Load

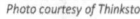

Photo courtesy of Thinkstock

NOTES

The center of gravity of a prefabricated section must be considered when loading and unloading flatbed trailers. If the load is not centered, the trailer may flip.

During construction of the Trans-Alaskan Pipeline, after inexperienced workers were crushed while loading or unloading trucks, ALYESKA required all new hires to attend several days of safety training in Anchorage before they were allowed on the job site.

Who has worked in manufacturing or construction? Do you want to tell us about it? What sort of safety training did you receive? What personal protective equipment was required?

SOURCE

Photo courtesy of Thinkstock

Inspect Chokers Prior to Lift

Photo courtesy of Thinkstock

NOTES

All chokers must be inspected prior to lift. If one of these should break, the load will shift and may cause injuries and property damage.

SOURCE

Photo courtesy of Thinkstock

Excavations

ARCHITECTURAL DESIGN AND CONSTRUCTION

Excavations

NOTES

A good foundation is essential. Let's talk about excavations.

Excavation

- U.S. Bureau of Labor Statistics (BLS) data show that 271 workers died in trenching or excavation cave-ins from 2000 through 2006 [BLS 2003-2009]

- Project designers have a role to play in excavation safety.

NOTES

Look at these statistics: over a 7-year period, 271 workers died in cave-ins, buried under tons of dirt. Cave-ins can be avoided. Use shoring or caissons [www.cdc.gov/niosh/topics/trenching/].

Contractors are responsible for site safety and often design formwork and other temporary structures. A best practice is to include a list of applicable safety practices associated with these temporary structures.

On some job sites, trenches can be sloped to grade. Depending on the soil report and the weather, temporary shores may be needed. A contractor who assumes that the trenches can be sloped to grade and then discovers the need for shores or piles could go bankrupt.

SOURCE

BLS [2003–2009]. Census of fatal occupational injuries. Washington, DC: U.S. Department of Labor, Bureau of Labor Statistics [www.bls.gov/iif/oshcfoi1.htm].

Wet Conditions Increase Risk

Photos courtesy of Thinkstock

NOTES

The angle of repose for loose earth varies from 30° to 45°. The backhoe operator is working close to the edge. How much rain would destabilize the slope under the backhoe? In the picture on the right, the dump truck overturned as it unloaded rock. What may have happened?

SOURCE

Photos courtesy of Thinkstock

Excavation Case Study

 **Excavation Court Case**

- Supreme Court of Mississippi

    - The heirs of a construction worker sued the project architects and others.

    - The worker and two others were killed when the walls of a ditch being excavated for a sewer line caved in, burying and smothering them.

Wanda M. Jones vs. James Reeves Construction,
93-CA-01139-SCT 9/20/1993 http://caselaw.findlaw.com/ms-supreme-court/1046041.html

NOTES

Before class, students should read the referenced Mississippi case. In class, discuss (1) the court's decision to not hold the designer liable and (2) the dissenting judge's opinion and how it could influence future cases.

SOURCE

FindLaw [1997]. Supreme Court of Mississippi. Wanda M. Jones vs. James Reeves Contractors Inc. Case no. 93-CA-01139-SCT. March 27 [caselaw.findlaw.com/ms-supreme-court/1046041.html].

Electrocution Case Study

 **Driller's Helper Electrocuted**

Safety tips to live by:

1. Watch for overhead dangers
2. Be aware of your surroundings
3. Know the machine capacity
4. Always secure loads
5. Drive safely
6. Be safe and smart

Alaska FACE Investigation 99AK019 www.cdc.gov/niosh/face/stateface/ak/99ak019.html

NOTES

Overhead power lines present a hazard for cranes, dump trucks, drilling equipment, concrete pumpers, and other equipment. Design considerations include the following:

- Indicating the locations of existing utilities on the contract drawings and marking a clear zone around the utilities
- Noting on the drawings the source of information and level of certainty on the location of underground utilities
- A plan for rerouting the power lines around the project site before construction begins

Safety tips to live by:

1. Watch for overhead dangers such as power lines and low structures
2. Be aware of your surroundings

3. Know load capacity of the machine
4. Always secure any loads that could slip or roll from forks
5. Drive safely
6. Be safe and smart

Students should read the NIOSH Fatality Assessment and Control Evaluation (FACE) report before coming to class. During the class discussion, students should be able to identify contributing factors and discuss their thoughts regarding site safety. Who was legally liable? What can the designer do to prevent this type of accident? What can the architect do about underground electrical cables and other utilities?

SOURCES

NIOSH FACE reports can be found at www.cdc.gov/niosh/face/.

NIOSH Fatality Assessment and Control Evaluation (FACE) Program [2000]. Driller's helper electrocuted when mast of drill rig contacted overhead power lines. Alaska FACE Investigation 99AK019 [www.cdc.gov/niosh/face/stateface/ak/99ak019.html].

Fatal Facts Accident Reports Index [www.setonresourcecenter.com/MSDS_Hazcom/FatalFacts/index.htm].

Photo courtesy of Alaska Division of Public Health

Alaska FACE Investigation 99AK019
July 12, 2000

Driller's Helper Electrocuted When Mast of Drill Rig Contacted Overhead Power Lines

SUMMARY

On June 17, 1999, a 32-year-old male drill truck operator's helper (the victim) was electrocuted when the mast of a drill rig contacted two-7,200-volt overhead power lines. The victim was assisting a drill rig operator to drill for a local environmental engineering contractor. They had relocated the truck to the front of an industrial lot to drill the last hole. A small flag indicating the well's position marked the location. The marker was near a fence separating the lot from an adjacent road. Above the marker were four power lines that ran parallel to the road. After extending the truck's front outrigger, the operator began raising the drill rig mast to position it over the marker. The victim was standing near the rear of the driver's side of the truck unloading equipment when the mast contacted the high voltage power line. Two workers employed by the contractor were standing several feet from the driver's side of the truck and heard a noise. They saw the victim and the operator frozen to and then collapse away from the truck. One worker went into a nearby building to call 911 as the other worker went to check both men. A worker from the building and a passerby arrived at the site as the first worker returned. Two teams were coordinated and CPR was started on the victim and operator. Emergency medical services arrived minutes later. The victim and the operator were transported to a nearby medical center where the victim was pronounced dead. The operator survived, but was unable to recall details of the incident.

Based on the findings of the investigation, to prevent similar occurrences, employers should:

- **Ensure that a hazard assessment has been completed to identify all hazardous conditions that may affect operation of equipment;**
- Ensure that equipment is not operated where any part is within 20 feet of electric power lines unless the lines have been de-energized and either grounded or insulation barriers have been installed;
- Ensure that a safety checklist is included in the written standard operating procedures (SOP) and is used prior to the start of any drill activity for each work site;

In addition, all companies responsible for marking drill sites should:

- **Maintain a minimum 20-foot safety zone and be knowledgeable of all applicable OSHA requirements for work near utilities and electric power supplies;**
- **Communicate to all parties involved in drilling activities the location of both above and below ground utility and electric power supplies near a drill marker that are within a distance equal to the height or extension of the drill equipment plus 20 feet.**

To help prevent or reduce the severity of injury in emergency situations, all drill rig owners and operators should:

- **Ensure that all operator's controls are in good working condition and are clearly labeled.**

INTRODUCTION

At approximately 2:00 PM on June 17, 1999, a 32-year-old driller's helper (the victim) was electrocuted when the mast of a drill truck contacted two-7,200-volt overhead power lines. On June 18, 1999, Alaska Department of Labor (AKDOL) notified the Alaska Division of Public Health, Section of Epidemiology. An investigation involving an injury prevention specialist for the Alaska Department of Health and Social Services, Section of Epidemiology, ensued on June 18, 1999. An on-site investigation was conducted on June 21, 1999. The incident was reviewed with AKDOL officials and the company owner. [It should be noted that due to injuries from the incident, the owner could not remember events proceeding, during, and following the incident.] Local police department, Alaska Medical Examiner, and AKDOL reports were requested.

The drilling operation in the incident was privately owned and operated. The company had been in business for approximately 10 years. The operator/owner had 25 years of drilling experience and was currently the sole drill operator (henceforth referred to as "the operator"). Normally, the company employed one permanent full-time helper to assist the operator in the shop and at drill sites; however, the company would employ one or more helpers and an additional operator depending on the work schedule and a project's specifications and timeline.

The victim had worked for the company for approximately 1½ years as a driller's helper and was training to become an operator. As a driller's helper, his responsibilities included driving the drill rig to and from a work site, equipment set-up (except raising the mast and leveling the drill rig), and facilitating drilling activities by retrieving and removing materials and supplies. In addition to these activities, as a trainee, the victim was taught drilling mechanics, general rock formations and characteristics, and use of the controls under the supervision of the operator. The day prior to the incident, the victim had operated the controls of the drill while being supervised by the operator. The victim was not permitted to operate the controls to raise the mast.

The company did not have a written safety program. Prior to drilling, the operator conducted daily tailgate or site specific safety meetings. During the tailgate safety meetings, any representatives or employees of a contractor working with the company during drilling operations were included. Topics usually included safe working practices, emergency stop procedures, and potential work site hazards. It could not be determined if a tailgate meeting was conducted between the operator and the victim; no records of these meetings were kept. However, tailgate meetings were a common practice during previous drill activities performed for this contractor.

The contractor involved in this incident was an environmental engineering agency with nearly 700 employees nationwide; 10 employees were on-staff at the incident location. The contractor had engaged the company for several projects during the 4 years prior to the incident. The contractor had a written safety program that included drill rig safety and personnel safety for its employees who were required to work with drilling companies.

INVESTIGATION

The company was engaged by the contractor to drill three monitoring wells for the placement of equipment to collect ground water data. The incident site was an industrial lot at the corner of two unpaved roads. Two buildings were on the lot. The lot surface consisted of compacted dirt, crushed rock, and gravel. A ditch and a 6-foot chain-link fence separated the front of the lot from the road. The fence continued around the perimeter of the lot. Access to the lot was over a culvert and through an 18-foot chain link gate. An electric utility easement was located along the front of the lot parallel with the road. Poles supporting four lines were located outside of the lot; the nearest utility pole was near the intersection of the two roads. The power lines were parallel to the road, and all lines (#1, #2, #3, and #4) traversed above the lot. Lines #1, #3, and #4 were energized to 7,200 volts; line #2 was neutral. A crossbeam with 3-foot spacing between lines supported the lines. The horizontal distance from line #4 (closest to the road) to the fence was approximately 11½ feet (Figure 1). Weather may have been a contributing factor at the time of the incident; reduced visibility from rain and mild mist was reported. The lot surface was very wet with some puddling of water.

Figure 1. Location of van and drill rig under overhead power lines

The equipment involved in the incident was a 1979 auger (mechanical rotary drill) mounted to a truck bed (Figure 2a and 2b). The truck was equipped with three outriggers: one at the front, mid-distance from both side; and two at the rear. The master derrick (or mast) was approximately 27 feet long with a maximum vertical rotation of 90 degrees from a rear pivot point 3½ feet from the end of the mast. The mast's pivot point was 8 feet above ground level. The height of the mast when fully raised was 31½ feet. All mast, drill, and pump controls were located at the rear of the truck. The control panel was in poor condition with worn and weathered labels (Figure 2c). An operator's platform consisted of a small metal grate attached to the rear of the truck below the control panel (Figure 2d).

Figure 2a. View of drill rig, front driver's side

Figure 2b. View of drill rig, passenger side

Figure 2c. Control Panel

Figure 2d. Control Panel and operator's platform

The procedure normally used by the drill operator and helper for setting-up the drill rig was as follows--

1. Extend and lower the front outrigger;
2. Raise the mast into position;
3. Extend and lower the rear outriggers;
4. Attach drill steel;
5. Check the drill rig stability (using Kelly bar) and level using the rear outriggers.

On the day of the incident, the company was drilling three monitoring wells. The company was responsible for drilling the holes, installing casing, and the sand pack. The contractor was responsible for requesting utility locations, marking the locations of the wells, and the collection of samples. Two employees of the contractor were working with the drill company to assure proper location of the wells, answer questions about a well location, and collect and process samples. Normally, only one employee of the contractor worked at the drill site; however, during this project, a second, more experienced employee was assisting due to the limited drill experience of the first. The employees had discussed proposed drill locations with the project manager and underground utility locates were done. The first two well locations were at the back of the lot and marked with flags. Neither employee was aware of the precise location of the third flag at the front of the lot until the day of the incident. Due to potential interference with a underground electric hook-up from the electric meter to one of the two buildings, the project manager placed the third flag after discussing the location with building occupants. [Secondary underground electric utility locates (past the meter's location) are not done by public utilities. A private electrical contractor must be hired to perform these locates.]

Drilling was briefly delayed due to the late arrival of the drill rig, but the activity was still on schedule. No other delays or problems had occurred while drilling the first two holes at the back of the lot. The truck was relocated to the front of the lot to drill the last hole. The marker was by the fence at the edge of the lot, and truck was backed into position under the overhead power lines. The truck was parked at a 90-degree angle (perpendicular) to the fence and the power lines. The back of the truck was 7 feet from the fence. The distance from the ground (at the truck's location) to the overhead lines was 31 feet. A van owned by the contractor was parked parallel to the truck, approximately 12 feet from the truck's driver side.

The contractor's employees (the witnesses) were standing near the gate, several feet from the van, discussing the location of the marker near the fence and potential involvement of underground utilities. The victim was removing equipment from the driver's side of the truck while the operator stood at the control panel at the rear of the truck. The front outrigger was extended and lowered. Neither of the witnesses was aware of the operator's activities and did not see the mast rise. The top of the mast rose between line #2 and #3, pushed line #3 toward the street, and contacted line #4. At this time, the witnesses heard a loud boom and crackle. Looking toward the drill rig, they saw sparks coming from the truck. Both the victim and the operator were "frozen" to the truck and then fell away from it toward the van. One witness instructed the other to go into a nearby building and call 911. Upon her return, two more people, both trained in first aid and CPR, joined them. They moved the victim and the operator to the other side of the van. CPR was started on both the victim and the drill rig operator. Emergency medical service personnel were dispatched at 2:05 PM, arriving at the site approximately 5 minutes later and continued CPR. The victim and the operator were transported to a nearby medical center where the victim was pronounced dead. The operator survived the event and was discharged 4 days later.

CAUSE OF DEATH

The medical examiner's report listed the cause of death as high voltage electrocution.

RECOMMENDATIONS/DISCUSSION

Recommendation #1: Employers should ensure that a hazard assessment has been completed to identify all hazardous conditions that may affect operation of equipment.

Discussion: In this case, the operator apparently did not consider the height of the mast when assessing for hazards during operation of the drill rig. Although the height of the power lines above the drill site marker may not have been known, recognition of a potential hazard would have prevented the incident.

Recommendation #2: Employers should ensure that equipment is not operated where any part is within 20 feet of electric power lines unless the lines have been de-energized and either grounded or insulation barriers have been installed.

Discussion: In accordance with 29 CFR 1926.416(g)(2)(iii)(A), "Any vehicle or mechanical equipment capable of having parts of its structure elevated near energized overhead lines shall be operated so that a clearance of 10 feet is maintained." While the truck was more than 20 feet below the overhead power lines, the mast entered the safety zone as it raised to a vertical position. Operators should maintain continuous sight of all potentially energized power sources when raising the mast to a vertical position. Employers should consider mandating a 20-foot safety zone; this has been adopted by many American and Canadian equipment manufacturers, professional trade groups, and operating engineer associations.

Recommendation #3: Employers should ensure that a safety checklist is included in the written standard operating procedures (SOP) and is used prior to the start of any drill activity for each work site.

Discussion: In this incident, workers either were not aware of or thought that the overhead power lines were an "adequate" distance from the drill rig. Overhead power lines and other potentially harmful conditions (e.g., pipe or rebar protruding from the ground) are not marked on the ground during locates. A person who is capable of identifying existing and predictable hazards at the work area or working conditions that are hazardous or dangerous must do a visual inspection of the work area prior to the start of all drilling activities. A site inspection checklist is an effective assessment tool to ensure that unsafe activities are avoided.

Recommendation #4: Companies responsible for selecting and marking drill sites should maintain a minimum 20-foot safety zone around utilities and electric power supplies and be knowledgeable of all applicable OSHA requirements for work near utilities and electric power supplies.

Discussion: To minimize the possibility of utility and electric power supply involvement at a work site, companies should develop and implement guidelines for marking drill sites that include potential hazards and possible alternatives and solutions for drill site selection when within 20 feet of utilities or electric power lines. Guidelines should emphasize a **20-foot safety zone around the utility or electric power supply**; additional space or clearance may be necessary to maintain a safety zone when equipment is raised.

Recommendation #5: Companies responsible for selecting and marking drill sites should communicate to all parties involved in drilling activities the location of both above and below ground utility and electric power supplies near a drill marker that are within a distance equal to the height or extension of the drill equipment plus 20 feet.

Discussion: The ability to communicate is crucial to safe work practices. Regardless of personal interpretation of how apparent or significant a hazard may be, information about **all** potential hazards and dangers at a work site should be communicated to workers.

In addition, to help prevent or reduce the severity of injury in emergency situations:

Recommendation #6: Drill rig owners and operators should ensure that all operator controls are in good working condition and are clearly labeled.

Discussion: Manufacturers recognized the increased risk for fatal injury when operator controls are not clearly labeled. Employers should ensure that all operator controls are clearly labeled before placing equipment in use. In the event of an emergency, basic operation or an emergency stop can be done.

REFERENCES

Clapp AL, Ed., National Electrical Safety Code Book, Fourth Ed. The Institute of Electrical and Electronics Engineers, Inc., 1996.

Office of the Federal Register: Code of Federal Regulations, Labor, 29 Part 1926. Washington, DC: U.S. Government Printing Office, 1999.

National Electrical Safety Code, C2-1997. The Institute of Electrical and Electronics Engineers, Inc., 1996.

Building Elements

ARCHITECTURAL DESIGN AND CONSTRUCTION
Building Elements

NOTES

This section will provide some examples of building elements for which design can influence construction safety. This is by no means a comprehensive list, nor is each discussion a comprehensive summary of the specific element and its associated potential hazards and influences. Each project is unique. The goal of this section is to learn more and then do more.

 **Roofs**

- Falls are the leading cause of fatal injuries and the second most common cause of nonfatal injuries in construction.

- In 2005, falls caused
    - 396 (32%) of 1,243 work-related deaths from injuries
    - 36,360 nonfatal injuries (23% were "lost time" accidents)

- One-third of the fatal falls were from roof edges or through holes [BLS 2003-2009]

NOTES

Does anyone know what a "lost time" accident is? A "lost time" accident requires the employee to miss one or more days of work. These accidents cannot be fixed with a Band-aid and an aspirin. Give me some examples.

One third of fatal falls were from roof edges or through holes. What can you do to prevent falls? How are commercial roofs different from your common house roof?

Roof hazards must be considered during both construction and maintenance. Some people who will be on the roof for maintenance will have nothing to do with roof maintenance—e.g., window washers, plumbers working on HVAC, etc.

SOURCE

BLS [2003–2009]. Census of fatal occupational injuries. Washington, DC: U.S. Department of Labor, Bureau of Labor Statistics [www.bls.gov/iif/oshcfoi1.htm].

Roof Hazards

- Access
- Fall from height
- Falling objects
- Heat/cold stress
- Material handling
- Structural collapse

Photo courtesy of T.J. Lyons

NOTES

We just discussed falls from a roof. How does the worker reach the roof? What might cause a roof to collapse? Material handling— what's that? What types of materials are used on a typical house roof? How do they get on the roof? Now, how is a commercial roof different? Can anyone tell me what kinds of equipment might be located on a roof? Do you know how much a chiller weighs? Are skylights heavy? Can anyone describe a parapet? What hazards are anticipated when handling solar panels?

SOURCE

Photo courtesy of T.J. Lyons

Methods to Reduce Roof Hazards

- Use parapets as guardrails
- Guardrail systems
- Anchor points
- Lifeline systems
- Prefabrication

Photo courtesy of T.J. Lyons

NOTES

Designers can enhance roof safety by including 39"- to 42"-high parapets at the roof's edge. These can be utilized during construction, if sequenced in early, and serve as fall protection for maintenance and construction workers who need to access equipment on the roof, perform maintenance, or replace the roof.

Another design consideration for renovation and demolition work is to review the condition and integrity of the existing structure and indicate any known hazards or deficiencies on the contract drawings.

Design appropriate and permanent roof fall protection systems to be used for construction and maintenance purposes. Consider permanent anchorage points, lifeline attachments, and/or holes in the perimeter for guardrail attachment.

As shown in the drawing, prefabrication of roof structures can reduce the need to work at height. One contractor has eliminated the construction of roofs on his projects by assembling them instead on the ground. This not only eliminates the potential for someone to fall but the contractor noted "we actually did it 30% faster than expected" since materials were easier to access and workers were not limited to the work area of one ladder step. He is now installing the shingles, lighting, ventilation ductwork, and electrical systems in the ceiling spaces, before the roof is lifted, further increasing safety and the speed of the work. This is Prevention through Design at its best.

SOURCE

Photo and prefabrication example courtesy of T.J. Lyons.

Parapets

The parapet will serve as adequate fall protection if it is at least 39" high.

Photo courtesy of Mike Behm

NOTES

Notice the parapet around this green roof. A safe walkway is delineated by pavers to the left of the plant bed.

SOURCE

Photo courtesy of Mike Behm

Railings Prevent Falls

Photo courtesy of Thinkstock

NOTES

This tank roof is surrounded by guard rails and there is a cage on the ladder. Schedule the installation of handrails, guardrails, and stair rails during the erection process to increase site safety. Provide pavers to identify safe walkways on fragile roofs.

SOURCE

Photo courtesy of Thinkstock

(PtD) Anchor Points

- Part of the facility

- Use during construction and maintenance

- OSHA standard regarding anchorages can be found in 29 CFR 1926.502(d)(15)

Photo courtesy of Thinkstock

NOTES

The worker is tied off to a safety line and to an anchor point.

The OSHA standard regarding anchorages can be found in 29 CFR 1926.502(d)(15):

1926.502(d)(15)

Anchorages used for attachment of personal fall arrest equipment shall be independent of any anchorage being used to support or suspend platforms and capable of supporting at least 5,000 pounds (22.2 kN) per employee attached, or shall be designed, installed, and used as follows:

1926.502(d)(15)(i)

as part of a complete personal fall arrest system which maintains a safety factor of at least two; and

1926.502(d)(15)(ii)

under the supervision of a qualified person.

If anchor points are designed into the building's structure, they can be used to facilitate safe window cleaning, roof repair, and other maintenance. If the anchor point can be sequenced in as early as possible, it can also be utilized during construction.

SOURCE
Photo courtesy of Thinkstock

Is this safe?

Photo courtesy of Thinkstock

NOTES

This worker is "tied off." Designers can provide an opportunity for construction workers to work safely, but ultimately it is up to workers to follow through and utilize safety elements. Look closely to see that the worker is wearing a fall protection harness.

SOURCE

Photo courtesy of Thinkstock

Fragile Roof Case Study

Walkways on Roof

Fragile roofing poses hazards to workers who need rooftop access

Electrician Dies Following a 60-foot Fall Through a Roof—Virginia, FACE 9605
www.cdc.gov/niosh/face/
In-house/full9605.html

Walkway guardrails designed as a barrier from fragile materials

Photo courtesy of the Virginia Division of Safety Research

NOTES

Students should read the case study before class and come to class prepared to discuss it.

Fragile roofing poses hazards to workers who need to access areas near the fragile pieces or who must walk past the fragile area to rooftop equipment. The following is a summary of a NIOSH fatality investigation (96-05), which demonstrates the need for designers to plan for permanent access around fragile roof elements.

A 21-year-old male electrician (the victim) died of injuries sustained from falling 60 feet through a roof. The victim and his apprentice co-worker had been dispatched to a locomotive repair building to work on electrical equipment on the roof. The two arrived at the job site about 1 p.m. and proceeded to the roof. The victim reportedly told the co-worker to follow in his footsteps, because there were numerous, barely distinguishable fiberglass panels on the rooftop. The victim walked down the slightly pitched roof to the ventilator, where the electrical work was to be performed. As he stepped around to the opposite side of the ventilator, he unintentionally stepped on a corrugated fiberglass roof panel. The panel broke, causing the victim to fall through the roof and strike the concrete floor, 60 feet below.

Even though the victim was aware of the hazard, he still stepped on the fiberglass panels because they were indistinguishable. Design can indeed influence actions and behaviors.

SOURCES

NIOSH Fatality Assessment and Control Evaluation (FACE) Program [1996]. Electrician dies following a 60-foot fall through a roof—Virginia. FACE 9605 [www.cdc.gov/niosh/face/In-house/full9605.html].

FACE report courtesy of Virginia Division of Safety Research

Photo courtesy of the Virginia Division of Safety Research

FACE 9605

Electrician Dies Following a 60-foot Fall Through a Roof--Virginia

SUMMARY

A 21-year-old male electrician (the victim) died of injuries received after falling 60 feet through a roof. The victim and his apprentice co-worker were dispatched to a locomotive repair building to repair electrical equipment located on the roof of the building. The two workers arrived at the job site about 1 p.m. and proceeded to the roof of the locomotive repair building. Once on the roof, the victim reportedly told the co-worker to follow in his foot steps since there were numerous, barely distinguishable fiberglass roof panels located on the roof top. The victim walked down the slightly pitched roof to the ventilator where electrical work was to be performed. The victim then walked around to the opposite side of the ventilator and unintentionally stepped on a corrugated fiberglass roof panel. The roof panel broke, causing the victim to fall through the roof and strike the concrete floor, 60 feet below. Two other employees, who were installing lighting fixtures inside the building, saw the victim fall through the air and strike the concrete floor. One worker rushed to the victim=s aid and checked for vital signs while the other worker called 911 for assistance. Cardiopulmonary resuscitation was performed by the worker until paramedics arrived. A medical evacuation helicopter arrived about 15 minutes after being called and transported the critically injured victim to a local hospital. The victim was pronounced brain dead about 43 hours after the incident occurred. NIOSH investigators concluded that, to prevent similar occurrences, employers should:

- *evaluate their current safety program and incorporate specific training procedures emphasizing the importance of recognizing and avoiding hazards in the workplace. These procedures should include, but not be limited to, conducting hazard evaluations before initiating work at a job site and implementing appropriate controls*
- *owners of buildings should evaluate and identify areas that may be hazardous to any personnel, including contractors, and prohibit access to these areas, or eliminate the hazard prior to access.*

INTRODUCTION

On October 25, 1995, a 21-year-old male electrician (the victim) died after falling 60 feet through a roof. On October 30, 1995, officials of the Virginia Occupational Safety and Health Administration (VOSH) notified the Division of Safety Research (DSR) of this fatality and requested technical assistance. On December 14, 1995, a safety specialist from DSR investigated the incident and reviewed the circumstances with the company owner, a manager at the locomotive repair building, and the VOSH compliance officer assigned to the case. Photographs of the incident site were obtained and the medical examiners report was reviewed.

The employer in this incident was an electrical contractor that had been in operation for 222 years and employed 12 workers, 4 of whom were electricians. The employer had a written general safety program and on-job-training was provided to all employees. Electricians were responsible for the enforcement of the safety program and they also conducted tail-gate safety meetings. The victim worked for the company for 5 years and 2 months prior to the incident. This was the first fatality the company had experienced.

INVESTIGATION

One of the employer's current contracts was to perform various electrical installation and repair services at a locomotive repair building. The contract had been ongoing for several years. The locomotive repair building was about 700-feet long by 170-feet wide by 80-feet high and was constructed in 1969. The roofing materials consisted mainly of corrugated metal panels with corrugated fiberglass panels interspersed into the roof at irregular intervals. Metal panels have structural integrity to support weight, whereas fiberglass panels do not. The corrugated fiberglass roof panels comprised at least 10 percent of all the roof panels and were faded/bleached from exposure to the weather thus looked similar to the metal panels. Additionally, 24 ventilators equipped with electric motors were located on the roof in a single line at the north end of the building.

On the day of the incident, the victim and his apprentice co-worker were assigned a job at a different location from where the incident occurred. Early in the afternoon the victim and co-worker were dispatched to the locomotive repair building to repair damages to an electric motor and wiring at one of the ventilators. The motor and wiring had been damaged in a fire the previous week. The two workers arrived at the building about 1 p.m. and climbed a fixed ladder on the outside of the building to the roof top. Once on the roof, the victim reportedly told the co-worker to follow in his footsteps, since there were numerous fiberglass roof panels all over the roof top. The two workers proceeded down the roof (pitch about 4:12) toward the damaged ventilator motor. Once at the ventilator the victim proceeded to the opposite side of the ventilator while the co-worker remained stationary. As the victim stepped around the ventilator and out of sight of the co-worker, he unintentionally stepped on a corrugated fiberglass roof panel (Figure). The panel broke and the victim fell through the roof to a concrete floor, 60 feet below. Two other company employees, who were installing lighting fixtures inside the building, saw the victim fall through the air and strike the concrete floor. One worker rushed to the victim=s aid and checked for vital signs while the other worker called 911 for assistance. Cardiopulmonary resuscitation was performed by one worker until paramedics arrived about 10 minutes after being called. A medical evacuation helicopter was summoned and arrived about 15 minutes later and transported the critically injured victim to a local hospital. The victim was pronounced brain dead about 43 hours after the incident occurred.

CAUSE OF DEATH

The medical examiner's report listed the cause of death as blunt-force head trauma.

RECOMMENDATIONS/DISCUSSION

Recommendation #1: Employers should evaluate their current safety program and incorporate specific training procedures emphasizing the importance of recognizing and avoiding hazards in the workplace. These procedures should include, but not be limited to, conducting hazard evaluations before initiating work at a job site, and implementing appropriate controls.

Discussion: The existence of a safety program is only the first step in obtaining a viable safety record. In addition to enforcement, safety programs should be evaluated and training procedures incorporated which emphasize the importance of recognizing and avoiding hazards in the workplace, following established safe work procedures, and wearing appropriate personal protective equipment. Before starting any work at a job site, the employer or employer's representative should identify, by observation and by collaboration with the job site owner, any potential or existing hazards. These hazards should be reviewed with the work crew, and methods to control the hazards and how to perform the work safely should be discussed. In this instance, the numerous irregularly spaced weathered corrugated-fiberglass roof panels could have been identified as a potential hazard because of their minimum load rating, proximity to the working area, and the visual similarity to the corrugated metal roof panels. The hazard of the corrugated fiberglass roof panels, although recognized by the victim, was not dealt with in an effective manner. Workers could have been instructed not to access the roof area until arrangements for safe access could be provided. Since the ventilators were all located in a single line across one end of the building, a walkway could have been constructed over the panels up to and around the ventilators for maintenance and repair. Alternatively, the corrugated fiberglass roof panels to and around the access area could have been replaced with metal corrugated panels, thus providing a stable walking/working surface, or a designated walkway marked with paint and protected by stanchions and handrails could have been installed.

Recommendation #2: Owners of buildings should evaluate and identify areas (e.g., roofs) that may be hazardous to any personnel, including contractors, and prohibit access to these areas, or eliminate the hazard prior to access.

Discussion: In 1969 metal and fiberglass corrugated roof panels were used in the construction of the roof of the locomotive repair building. Additionally, 26 ventilators equipped with electrical motors were installed on the roof, on one end of the building, to ventilate exhaust fumes from the locomotives. The fiberglass panels accounted for about 10% of all panels and were irregularly spaced among the metal panels. Also, the fiberglass panels were faded, due to weathering, and resembled the metal panels in appearance. These conditions should have been evaluated and appropriate action to mitigate the hazards should have been taken before access to the roof area was permitted.

Pascalは、例外...

Ladder Placement Case Study

United Kingdom CDM Case

The UK Construction Design and Management (CDM) regulations were discussed in the Overview module.

Construction Industry Research and Information Association (CIRIA) [2004]. CDM regulations: work sector guidance for designers. 2nd Ed. London: CIRIA.

Architect fined after health and safety lapse causes death
www.bdonline.co.uk/30-july-2010/20050.issue

NOTES

Students should read the case study before class and be prepared to discuss it. This information can help the instructor lead a discussion:

It is designers' responsibility, under CDM, to

- make sure that they themselves are competent and adequately resourced to address the health and safety issues likely to be involved in the design
- check that clients are aware of their duties
- avoid, during the design process, any foreseeable risks to those involved in the construction and future use of the structure. In doing so, they should eliminate hazards (so far as is reasonably practicable when taking into account other design considerations) and reduce risk associated with those hazards that remain

- provide adequate information about any significant risks associated with the design
- coordinate their work with that of others in order to improve the way in which risks are managed and controlled.

How can you design the building to facilitate safe maintenance? In this instance, would a parapet have saved a life? Appropriately designed parapets reduce the risk of falling. Consider the placement of the ladder relative to the opening. Could the ladder have been placed in a safer location?

SOURCE

Rogers D [2010]. Architect fined after health and safety lapse causes death. Building Design [bdonline] 30 July [www.bdonline.co.uk/30-july-2010/20050.issue].

(PtD) **Skylights**

In 2003, worker deaths included these falls:

- 23 through skylights

- 11 through existing roof openings

- 24 through existing floor openings

Most of these deaths occurred in the construction industry.
[BLS 2003–2009]

NOTES

Skylights are trendy. How can you design skylights for safety?

SOURCES

NIOSH [2004]. Preventing falls of workers through skylights and roof and floor openings. Cincinnati, OH: U.S. Department of Health and Human Services, Centers for Disease Control and Prevention, National Institute for Occupational Safety and Health, DHHS (NIOSH) Publication No. 2004–156.

BLS [2003–2009]. Census of fatal occupational injuries and current population survey [www.bls.gov/iif/oshcfoi1.htm].

Skylight Installation Fatality

Fatality During Skylight Installation

Photo courtesy of the California Department of Public Health

An Electrical Worker Dies When He Falls Through a Skylight While Installing Solar Panels on the Roof of a Warehouse www.cdc.gov/niosh/face/stateface/ca/09ca003.html

NOTES

Skylights are fragile roof openings. How can they be designed for safety during construction? Will that design protect workers during maintenance?

Suggestions include designing domed rather than flat skylights, with shatterproof glass or the addition of strengthening wires; and designing guardrail protection around skylights.

An American National Standards Institute (ANSI) committee is developing a standard that will require skylight manufacturers to meet certain safety requirements on their products. Designers must consider the impact resistance of skylights specified.

SOURCES

NIOSH Fatality Assessment and Control Evaluation (FACE) Program [2009]. An electrical worker dies when he falls through a skylight while installing solar panels on the roof of a warehouse. California case report no. 09CA003 [www.cdc.gov/niosh/face/stateface/ca/09ca003.html].

Photo courtesy of California Department of Public Health

Fatality Assessment and Control Evaluation (FACE) Program

An Electrical Worker Dies When He Falls Through a Skylight While Installing Solar Panels on the Roof of a Warehouse

California Case Report: 09CA003

California Case Report

Summary

AA 46-year-old electrical worker died when he fell through a skylight on a roof while installing solar panels. The victim was carrying solar panels and walking backwards because of the limited space around the skylight. As the victim was walking backwards, he tripped on the raised edge of the skylight frame and fell onto the skylight. The skylight glazing (the transparent portion of the skylight) broke under the impact and the victim fell approximately 40 feet to the ground below. Although the skylight label indicated that it was tested in accordance with OSHA fall protection standards, there are currently no uniform test criteria to determine material strength to withstand worker impact. The CA/FACE investigator determined that in order to prevent future incidents, employers of workers who install solar panels on roofs with skylights should:

- **Develop, implement, and enforce a fall protection program to prevent falls through skylights.**

Introduction

On Monday April 6, 2009, at approximately 2:00 p.m., a 46-year-old electrical worker died when he fell through a skylight approximately 40 feet to the ground below. He was installing solar panels on the roof of a warehouse when the incident occurred. The CA/FACE investigator was notified of this incident on May 8, 2009, from the Department of Investigations of the Division of Occupational Safety and Health (Cal/OSHA). On June 15, 2009, the CA/FACE investigator interviewed the risk manager of the electrical contractor that employed the victim. Telephone interviews were conducted on July 2, 2009, with the local union business manager, one co-worker of the victim, and a sales manager of the skylight manufacturer. A site visit was performed on July 29, 2009. The CA/FACE investigator reviewed the electrical contractor's safety policies and procedures, training program, and the victim's orientation and training records.

The victim was a member of the local electrical union with eight years of work experience. He had been employed with the electrical contractor for seven days when the incident occurred. The electrical contractor had been in business for 46 years and had 800 employees. There were 33 employees at the site the day the incident occurred. The employer had a complete written Injury and Illness Prevention Program (IIPP) which included sections on working from elevations, fall hazard awareness, and fall protection. They also had a written training program. The victim had received employee orientation and safety awareness training on the first day of hire. The victim also attended daily site-specific safety awareness tailgate meetings for the site at which they were working.

Back to Top

Investigation

The site of the incident was a flat roof of an active warehouse with a surface area of approximately 650,000 square feet. There were a total of 357 skylights on the roof. Alternate rows of skylights were sealed closed and had stamped ratings; the other rows had skylights that opened and had antitheft

NIOSH FACE Home

State-based Case Reports

California Case Reports

On This Page...

- Summary

- Introduction

- Investigation

- Cause of Death

- Recommendations and Discussion

- References

- Exhibits

- California FACE Program

grating underneath. This solar panel rooftop installation was designed to produce approximately one megawatt of electricity for the Southern California area. In order to do this, 16,272 solar panels were being installed on the vacant area of the roof between the rows of skylights. The foundations for the solar panels were in close proximity to the skylights and workers had approximately 18 inches of clearance to maneuver around the skylights as the system was being built. The workers also had to be aware of the weight restriction on the roof, as they were not allowed to exceed 300 pounds per square foot.

On the day of the incident, the victim and his co-worker were carrying and installing electrical solar panels on the roof of the warehouse. No fall protection was used as the skylights were marked as being in accordance with OSHA fall protection standards. The solar panels were boxed and placed on the roof by a crane. Each panel was approximately two feet wide by four feet long, and weighed 24 pounds. At approximately 2:00 p.m., the victim and co-worker were carrying two panels at a time. As they approached a skylight, they had to maneuver around it using the 18 inches of clearance. The victim turned and walked backwards, and tripped on the raised edge of the skylight. He landed on the skylight in a sitting position and then, without warning, the plastic dome glazing broke. As the victim started to fall, a co-worker reached out and tried to grab his foot, but was unable to reach him in time. The victim fell approximately 40 feet to the warehouse floor below. Numerous workers with radios immediately called the office to report the incident and those with cell phones immediately called 911. The paramedics and fire department responded within minutes and pronounced the victim dead at the scene.

Back to Top

Cause of Death

The cause of death according to the death certificate was multiple blunt force injuries.

Recommendations/Discussion

Employers who use machines that recycle waste products should ensure that:

Recommendation #1: Develop, implement, and enforce a fall protection program to prevent falls through skylights.

Discussion: In this particular incident, the skylight was marked by the manufacturer that it was "tested in accordance" with OSHA fall protection standards. According to the employer's risk manager, the general contractor had reviewed the job safety requirements including fall protection plans. Based on the information that the skylight had been tested in accordance with OSHA standards, no other fall protection measures were implemented on this job site. There are currently no uniform test criteria to determine material strength of skylights to withstand worker impact. Such test criteria would include the degradation of plastic or plastic containing materials after several years of sun exposure and the ability to withstand a point impact. One organization, ASTM International, is currently developing such testing guidelines. At this time, employers should not assume that manufacturer testing ensures that a particular skylight can sustain the impact of a worker. In order to prevent falls through skylights, employers should implement and maintain a fall protection program that includes:

- Skylight screens capable of safely supporting the greater of 400 pounds or twice the weight of the employees plus his equipment and materials, or
- Guardrails around the skylight at least 45 inches in height with a top rail and mid rail which should be half way between the bottom surface and top rail. The rails should be able to withstand a live load of 20 pounds per square foot.

If these two methods are not feasible, then the use of personal fall protection should be utilized. A personal fall protection system consists of a body harness, lanyard and anchor points. Had any of these fall protection methods been used at this job site, the victim would not have fallen through the skylight to the ground below.

http://www.cdc.gov/niosh/face/stateface/ca/09ca003.html[7/6/2012 3:00:39 PM]

References

General Industry Safety Orders Article 2. Standard Specifications. §3209. Standard Guardrails. §3212. Floor Openings, Floor Holes and Roofs. (b)(e) Subchapter 4. Construction Safety Orders

Construction Safety Orders Article 24. Fall Protection. §1632. Floor, Roof, and Wall Openings to Be Guarded. §1670. Personal Fall Arrest Systems, Personal Fall Restraint Systems and Positioning Devices.

http://www.energy.ca.gov/links/base.php?pagetype=solar (link not available)

http://www.aamanet.org/general.asp?sect=1&id=291

http://www.astm.org/DATABASE.CART/WORKITEMS/WK17797.htm

Back to Top

Exhibits

Exhibit 1. The rooftop looking south.

NIOSH FACE Program: California Case Report 09CA003 | CDC/NIOSH

Exhibit 2. The rooftop looking west.

Exhibit 3. The skylight the victim fell through.

NIOSH FACE Program: California Case Report 09CA003 | CDC/NIOSH

Exhibit 4. The solar panels that the victim was carrying when he tripped.

Exhibit 5. A view of the incident scene and the base foundations for the solar panels.

http://www.cdc.gov/niosh/face/stateface/ca/09ca003.html[7/6/2012 3:00:39 PM]

Exhibit 6. The identification tab on the skylight the victim fell through.

Back to Top

California Fatality Assessment and Control Evaluation (FACE) Project

The California Department of Public Health, in cooperation with the Public Health Institute and the National Institute for Occupational Safety and Health (NIOSH), conducts investigations of work-related fatalities. The goal of the CA/FACE program is to prevent fatal work injuries. CA/FACE aims to achieve this goal by studying the work environment, the worker, the task the worker was performing, the tools the worker was using, the energy exchange resulting in fatal injury, and the role of management in controlling how these factors interact. NIOSH-funded, state-based FACE programs include: California, Iowa, Kentucky, Massachusetts, Michigan, New Jersey, New York, Oregon, and Washington.

To contact California State FACE program personnel **regarding State-based FACE reports, please use information listed on the Contact Sheet on the NIOSH FACE web site. Please contact** In-house FACE program personnel **regarding In-house FACE reports and to gain assistance when State-FACE program personnel cannot be reached.**

◀ California Case Reports

Unguarded Skylight Fatality

Unguarded Flat Skylight

Photo courtesy of the Wisconsin Department of Public Health

Laborer Dies From Fall Through Skylight While Shoveling Snow on Roof
www.cdc.gov/niosh/face/stateface/wi/99WI002.html

NOTES

A laborer charged with clearing snow off of a roof fell through a skylight to his death. Properly designed protection of the skylight would have prevented the accident. Do we rely on telling employees to "be careful"? How can we solve this recognizable safety issue?

SOURCES

NIOSH Fatality Assessment and Control Evaluation (FACE) Program [1999]. Laborer dies from fall through skylight while shovelling snow on roof. Wisconsin case report no. 99WI002 [www.cdc.gov/niosh/face/stateface/wi/99WI002.html].

Photo courtesy of Wisconsin Department of Public Health

Slide 53

FACE 99WI00201

Laborer Dies From Fall Through Skylight While Shoveling Snow on Roof

SUMMARY: A 43-year-old male laborer (the victim) at a coatings manufacturing company died after falling through a skylight to a concrete floor 14 feet below. The victim and a co-worker had volunteered to clear snow from the roof of the company building late on a January afternoon, after their regular work day was finished. A flat roof over the first story portion of the company was covered with drifted snow, which varied in depth from several inches to over 3 feet in places. The victim was using large steps to walk through the snow to the south side of the roof, where the snow was deepest. It completely covered the tops of skylights in this area. He apparently failed to see the unguarded, three-foot square skylight and stepped onto it while walking. The plastic bubble of the skylight broke, and the victim fell to the concrete floor. His co-worker had been walking toward the north side of the roof, and turned to look when he heard a noise from the victim's direction. Seeing the broken skylight, the co-worker yelled for help. Workers on the main floor were already assisting the victim, where he had fallen, and called emergency services. EMS responders were on the scene within minutes, and a physician pronounced the victim dead at the scene. To prevent future fatalities of this type, the FACE investigator recommends employers should:

- **guard skylight openings**

- **lock all doors that provide access to unguarded rooftops.**

- **provide training in the recognition and avoidance of unsafe conditions to workers who are assigned tasks outside their normal duties.**

INTRODUCTION:

On January 5, 1999, a 43-year-old black male laborer died after falling 14 feet through a skylight opening. The Wisconsin FACE field investigator learned of the incident from a newspaper article on January 6, 1999. On March 24, 1999, the field investigator conducted a phone interview with the employer. The investigator viewed the site from the street on a later date, and obtained the death certificate and the medical examiner and sheriff's reports.

The employer was a plastic coatings manufacturer that had been in business about forty years. There have been no fatalities at the company prior to this incident. The victim had worked for the company for about eighteen years. His work duties usually involved mixing paint used in the plastics manufacturing process. He was characterized as a worker who frequently volunteered to help the company on overtime assignments and with chores that were outside of his regular duties. Workers received on-the-job training for their assigned jobs. Safety information was included with job training, and during company training sessions for updates on workplace requirements. Written safety plans were in place for routine job activities, but not for the activity of clearing snow from a roof.

INVESTIGATION:

The incident occurred at a plastic coating manufacturing factory. The facility had a two-story portion where the manufacturing processes occurred, and an attached one-story warehouse area. Two doors on the second story provided access to the flat roof of the first story. The employer had designated the doors as emergency exits for an event requiring evacuation of the manufacturing area. Employees often used the flat rooftop for breaks, using the emergency doors for access. The victim had used the rooftop for his breaks, but it is unknown if he had ever cleared snow from the roof before the day of the incident.

On the day of the incident, the victim worked his regular shift starting at 6:30 AM, with a lunch break. Near the end of the shift, his supervisor asked for help in clearing snow and ice from the front of the emergency exit doors. The victim volunteered to help, so at about 4:15 PM, the supervisor and the victim went to the roof. There was daylight at this time. The supervisor began walking toward the north end of the roof to use a snowblower, and the victim headed toward the south end. Footprints in the snow indicate the victim was taking long strides, and after walking approximately 100 feet he stepped onto a skylight that was buried in the snow. The light plastic covering of the skylight gave way, and he plunged through the 3-foot square roof opening to the concrete floor of the warehouse below. His supervisor turned to look in the direction of the victim when he heard a noise. Seeing the broken skylight, he yelled for help. Workers on the main floor were already assisting the victim, where he had fallen, and called emergency services. EMS responders were on the scene within minutes, and a physician pronounced the victim dead at the scene.

CAUSE OF DEATH: The medical examiner's report listed the cause of death as craniocerebral, neck and chest injuries due to a fall from height.

RECOMMENDATIONS/DISCUSSION

Recommendation #1: Employers should guard skylight openings.

Discussion: According to 29 CFR 1910.23 (a)(4) "Every skylight floor opening and hole shall be guarded by a standard skylight screen or a fixed standard railing on all exposed sides." Additional requirements are specific about the design and capacity of the guard. In this case, the plastic bubble shell over the skylight opening did not meet the guarding standard, and the victim fell through the opening when he accidentally stepped on the shell.

Recommendation #2: Employers should lock all doors that provide access to unguarded rooftops.

Discussion: The rooftop where the incident occurred was frequently used by employees for breaks. It had also been designated by the company as an exit area for emergency evacuation. The roof was not designed for these purposes, but the company did not prohibit access to the rooftop. Entrance to hazardous areas through doorways should be controlled with doorlocks, and should be marked with warning signs. While not directly a cause of this fatality, the unsafe practice of using the rooftop for breaks placed the victim and co-workers at risk of falling through skylights and from the roof edge.

Recommendation #3: Employers should provide training in the recognition and avoidance of unsafe conditions to workers who are assigned tasks outside their normal duties.

Discussion: The victim's primary job responsibilities were in the plastics manufacturing process. Shoveling snow from a rooftop was outside of his normal duties, and presented hazards that may not have been immediately evident to the worker. When workers are expected to perform additional duties that place the worker at risk, the employer should instruct the worker on those risks and how to avoid them.

REFERENCES

29 CFR 1910.23 (a)(4) Code of Federal Regulations, Washington D.C.: U.S. Government Printing Office, Office of the Federal Register.

http://www.cdc.gov/niosh/face/stateface/wi/99WI002.html[7/6/2012 3:07:15 PM]

NIOSH FACE Program: Wisconsin Case Report 99WI002 | CDC/NIOSH

Figure 1. This picture shows the broken skylight where victim fell.

Figure 2. This picture shows an adjacent skylight, which was unguarded and snow covered.

http://www.cdc.gov/niosh/face/stateface/wi/99WI002.html[7/6/2012 3:07:15 PM]

Figure 3. This picture shows the rooftop view with snow drifts over the skylights.

FATAL ASSESSMENT AND CONTROL EVALUATION (FACE) PROGRAM

FACE 99WI00201

Staff members of the FACE Project of the Wisconsin Division of Public Health do FACE investigations when a work-related death is reported. The goal of these investigations is to prevent fatal work injuries in the future by studying: the working environment, the worker, the task the worker was performing, the tools the worker was using, the energy exchange resulting in fatal injury and the role of management in controlling how these factors interact.

To contact Wisconsin State FACE program personnel regarding State-based FACE reports, please use information listed on the Contact Sheet on the NIOSH FACE web site. Please contact In-house FACE program personnel regarding In-house FACE reports and to gain assistance when State-FACE program personnel cannot be reached.

Back to Wisconsin list

NIOSH FACEWeb

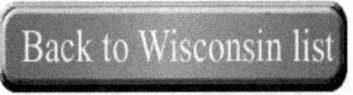

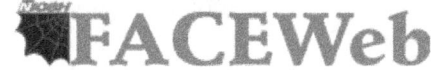

Skylight with Guard Cage

Photo courtesy of Plasteco

NOTES

The skylights in this picture are covered with a guard cage.

SOURCE

Photo courtesy of Plasteco

AC Unit Maintenance Fatality

 **AC Unit Maintenance**

- 2000
    - Renovation, addition to existing building
    - 12 existing skylights were located on lower roof
    - Several existing AC units located on lower roof
    - New AC units located on raised roof
    - One towards the edge of the raised roof
    - Roof is split level, ~8 meters
- 2002
    - Contractor hired to service air conditioning units

NOTES

Here is another case study. In 2002, a contractor was hired to service air conditioning units on the roof.

 **AC Unit Maintenance**

Consider:

1. Comparison with Mississippi case
2. Judgment against architect
3. Could this judgment happen in the U.S.?
4. Was the risk foreseeable?
5. Was the ruling fair?

Iannello v. BAE Automation and Electrical Services Pty Ltd & Ors
www.austlii.edu.au/au/cases/vic/VSC/2008/544.html

NOTES

Students should read the referenced case study and come prepared to discuss it in class.

SOURCE

AR Conolly & Company Lawyers. Iannello v. BAE Automation and Electrical Services Pty Ltd & Ors. Supreme Court of Victoria, VSC 54 [2008]. Benchmark Daily Bulletin Dec 9:3 [www.arconolly.com.au/benchmark/composite/benchmark_09-12-2008_insurance_banking_construction.pdf].

Sketch of Rooftop

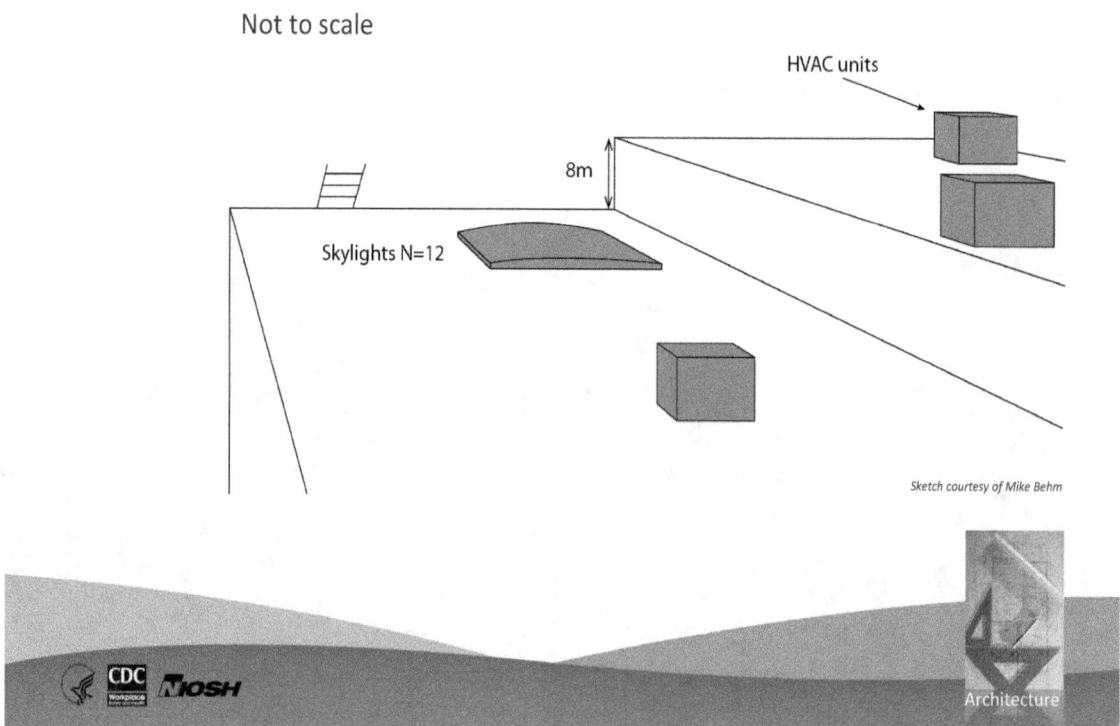

Not to scale

HVAC units

8m

Skylights N=12

Sketch courtesy of Mike Behm

NOTES

Here is a sketch of the roof. Was the worker just unlucky?

SOURCE

Sketch courtesy of Mike Behm

 **Green Roofs**

- Green (vegetated) roofs becoming popular in United States

- Present new hazards for landscapers and maintenance crews

[Luckett 2010]

Photo courtesy of Carol Clinton

NOTES

Green roofs are also trendy. The Center for Urban Greenery and Ecology (CUGE) in Singapore has developed guidelines entitled "Design for Safety on Rooftop Greenery," available at www.amazon.com/Design-For-Safety-Rooftop-Greenery/dp/9810852320.

Vegetated roofs are not "maintenance free"; frequent roof access is required to maintain vegetated roofs [Luckett 2010]. The installation and maintenance of vegetated roof materials present unique hazards and an increased risk to roofers and landscapers. What do you think those hazards might be?

One best practice for green roof maintenance is to post a document at the hatch or interior access point that shows the location of utilities and infrastructure on the roof, the profile of the green roof, and the list of warranties associated with the roof.

SOURCES

Luckett K [2010]. Green roof construction and maintenance. New York: McGraw Hill.

Photo courtesy of Carol Clinton

Green Roof Safety Design
[Weiler and Scholtz-Barth 2009]

Issues	Design Ideas
Access for people, tools, materials	Fixed stairs inside, designated walkways
Ergonomics	Allow adequate space to work. Include on-site storage for tools, fertilizers, etc.
Falls at building edge	Parapets, lifelines, anchorage systems
Falls in roof openings	Guard skylights and other roof openings
Fire, wind uplift	Vegetation-free zones
Maintenance	Plant-selection strategies
Rooftop machinery hazards	Machinery guards

NOTES

These issues and design ideas are a result of a study on hazards and risks associated with vegetated roofs. Green roofs need regular inspection and maintenance, so access for people, tools, and materials is typically provided via a fixed stair rather than a ladder. Designated walkways, often pavers, are used to access plant beds. Plant selection can minimize maintenance. Safety features to prevent falls apply to all roofs. Fire is a hazard on vegetated roofs and has occurred in dry areas and during drought. An ANSI standard has been developed to design green roofs for fire safety. A vegetation-free zone can be used to protect against wind uplift and fire. Rooftop machinery needs to be protected against windblown debris. Machine guards can protect workers and the machinery simultaneously.

SOURCES

American National Standards Institute (ANSI) [2010]. ANSI/SPRI VF-1, External Fire Design Standard for Vegetative Roofs [www.greenroofs.org/resources/ANSI_SPRI_VF_1_Extrernal_Fire_Design_Standard_for_Vegetative_Roofs_Jan_2010.pdf].

Weiler S, Scholtz-Barth K [2009]. Green roof systems: A guide to the planning, design and construction of building over structure. New York: John Wiley & Sons.

Safe Roof Garden

Garden rooftop patio with railings to prevent falls

Photo courtesy of Thinkstock

NOTES

This roof has a guardrail, planters, and a vine-covered trellis. Green roofs need hydration and thus water access for the plants. Keep in mind that hydration systems require maintenance and repair, and plantings require weeding.

SOURCE

Photo courtesy of Thinkstock

 **Unsafe Vegetated Roof**

NOTES

What could have been done differently from a design standpoint? Note that there is no parapet, no lifeline or anchorage system, and no permanent access to the roof.

SOURCE

Photo courtesy of Mike Behm

 **Installing Rails for Solar Panels**

How could this man work safer?

Photo courtesy of Thinkstock

NOTES

This worker is wearing a hardhat. What other PPE should he be using?

SOURCE

Photo courtesy of Thinkstock

Windows and Atria

How would you wash these
windows or replace a broken
pane?

Photo courtesy of Thinkstock

NOTES

Windows and atria present unique situations for installation and maintenance. The CIRIA document provides the following considerations for design.

Installation: Has access for installation been considered? Does the public need to be protected? Will scaffolding be required? How will the components be lifted and fixed in place?

Cleaning and maintenance: Has consideration been given to means of cleaning both inside and outside? How is access to be achieved for cleaning? Does it fit with the building's intended use? Will other services interfere with cleaning? What ergonomic issues should be considered?

SOURCE

Photo courtesy of Thinkstock

Unsafe Window Maintenance

Photo courtesy of Thinkstock

NOTES

Certainly, the employee in this picture should not have attempted this job. There is no place to tie off. Could anchorage points or a lifeline system be designed in place? Could an alternative, safer access be designed into the building? Consider potential design solutions. Could equipment, such as a bucket truck, be utilized?

What if the company in charge of maintenance told the building owner of the risks associated with maintaining these windows as designed and included the equipment rental in the bid? Engineering controls are less effective than a design solution because they make assumptions about downstream activities and decisions. The next slide shows an example to consider.

SOURCE

Photo courtesy of Thinkstock

Window Access System

Safe access for cleaning and maintenance of the facility should be considered during the design phase.

Photo courtesy of Thinkstock

NOTES

Providing safe access for cleaning windows and servicing equipment can be accomplished in the design phase of a project. Resist suggestions during a "value engineering" review to eliminate these features by providing a realistic estimate of the costs to rent or purchase special equipment to perform maintenance. This bucket lift allows workers access to all areas necessary to clean and maintain the glass.

SOURCE

Photo courtesy of Thinkstock

General Considerations

ARCHITECTURAL DESIGN AND CONSTRUCTION
General Considerations

NOTES

We've covered specific hazards associated with skylights, fragile roofs, and green roofs. Next we'll look at some general considerations.

(PtD) Material Handling

Heavy blocks are a significant musculoskeletal hazard, causing many injuries, but are an easy design issue to resolve.

Photo courtesy of Thinkstock

NOTES

In the United Kingdom, the Construction Industry Advisory Committee (CONIAC) has concluded that there is a high risk of injury in the single-handed, repetitive manual lifting of building blocks heavier than 20 kg, and this should be taken into account before specifying heavy units. This is easily dealt with in the specification or prohibited-materials list, and then it becomes a contractor's issue to manage the design specifications and ensure that they have been carried out.

SOURCES

Construction Industry Advisory Committee [CONIAC] [www.hse.gov.uk/guidance/index.htm]

Photo courtesy of Thinkstock

 **Surface Coatings and Finishes**

- Why apply?
- Must be sprayed?
- Materials compatible?
- Working space?
- Ventilation?
- Pretreat materials?
- Handling issues?
- Access issues?
- Is there a need for respiratory protection?

Photo courtesy of Thinkstock

This worker wears protection against finish hazards.

NOTES

Surface coatings and finishes can present health hazards to workers that include exposure to volatile organic compounds and isocynates and slick surfaces. Designers can mitigate the risk of these hazards. To what extent does the nature of the selected material and its requirements for application either reduce or cause hazards? Does the sequence of projected construction activities identify situations where unacceptable hazards might occur, such as application in poorly ventilated areas or confined spaces? Think about the entire building life cycle. How often is a space painted? What problems are associated with carpet? Do other floor finishes present problems?

SOURCE

Photo courtesy of Thinkstock

Building Decommissioning

ARCHITECTURAL DESIGN AND CONTRUCTION
Building Decommissioning

NOTES

Designers can play a role in worker safety and health during decommissioning and refurbishment. High-profile hazards such as asbestos and lead-based paint are still prevalent in older buildings. A complete investigation, including a site survey, would assist designers in their assessment of construction hazards.

PtD Demolition

DANGER
DEMOLITION
KEEP OUT

Photo courtesy of Thinkstock

NOTES

Designers should also consider the work that must be performed to demolish or refurbish a structure. Initiating a site assessment is a key first step. Specifically, these design suggestions should be considered.

1. Before demolishing and renovating any roof structure that is damaged, ensure that an engineering survey is performed by a competent person to determine the condition of the roof, trusses, purlins, and the structure itself. This survey would evaluate the stability of the structure and its components. The survey should suggest how fall-protection devices will be incorporated into the damaged structure.

2. Before demolishing and renovating any structure, ensure that an engineering survey is performed by a competent person to determine the condition of the structure, evaluate the possibility of unplanned collapse, and plan for potential hazards. These hazards include a high level of dust.

3. Identify potentially hazardous materials such as asbestos and lead paint and take precautions to minimize worker exposure to them.

SOURCE

Photo courtesy of Thinkstock

 **Refurbishment**

During remodeling, minimize risks to
Eyes: Safety glasses
Skin: Long sleeves, pants,
shoes and socks
Hands: Gloves
Ears: Earplugs
Head: Hardhat
Nose & Mouth: Face mask
Lungs: Exhaust fan

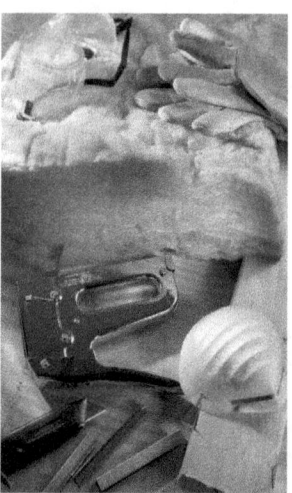

Photo courtesy of Thinkstock

NOTES

How many of you have installed fiberglass insulation? Did the insulation make your arms itch? Did you wear a face mask? Do you think you inhaled any fiberglass or plaster dust? Some of that may still be in your lungs. What could you do to minimize exposure to fiberglass and plaster particulates? Wear long sleeves, a face mask, eye protection, and gloves. Use an exhaust fan.

SOURCE

Photo courtesy of Thinkstock

Summary

 **Recap**

- **Prevention through Design (PtD)** is an emerging process for saving lives, time, and money and for protecting workers' health.

- PtD is the smart thing to do and the right thing to do.

- Although site safety is the contractor's responsibility, the designer has the ethical duty to create drawings with good constructability.

- There are tools and examples to facilitate PtD.

NOTES

PtD saves lives, time, and money and reduces or eliminates risks for workers. PtD is the ethical thing to do. Good constructability is the designer's responsibility.

Help make the workplace safer...

Include *Prevention through Design* **concepts in your projects.**

For more information, please contact the National Institute for Occupational Safety and Health (NIOSH) at

> **Telephone: (513) 533–8302**
> **E-mail:** preventionthroughdesign@cdc.gov

Visit these NIOSH Prevention through Design Web sites:

www.cdc.gov/niosh/topics/PtD/

www.cdc.gov/niosh/programs/PtDesign/

NOTES

This presentation was intended to provide examples of construction hazards and risks that could be positively or negatively affected by design decisions. It is certainly not comprehensive in any way. All members of the construction project team (owner, designers, contractors, and safety professionals) must attempt to learn more about construction site safety early in the built environment's life cycle. The earlier more is learned, the more effective and safer the process can be. Each party has a role to play. The United Kingdom and Australia have promulgated designers' roles and responsibilities for safe construction design. Those designers are still learning how to identify and manage risks and how they can provide safer and healthier designs. We encourage the infusion of construction and safety knowledge into the design team and design reviews. Organizations and individuals seeking to positively impact construction workers' safety and health through design will need first an open mind and second a holistic view of what factors influence workers' actions and inactions. Are there any questions?

SOURCES

NIOSH Prevention through Design Web sites:

www.cdc.gov/niosh/topics/PtD/

www.cdc.gov/niosh/programs/PtDesign/

References

American Institute of Industrial Hygienists [AIHA] [2008]. Strategy to demonstrate the value of industrial hygiene [www.aiha.org/votp_NEW/pdf/votp_exec_summary.pdf].

ANSI/AIHA [2005]. American national standard for occupational health and safety management systems. New York: American National Standards Institute, Inc. ANSI/AIHA Z10-2005.

AR Conolly & Company Lawyers. Iannello v. BAE Automation and Electrical Services Pty Ltd & Ors. Supreme Court of Victoria, VSC 54 [2008]. Benchmark Daily Bulletin Dec 9:3 [www.arconolly.com.au/benchmark/composite/benchmark_09-12-2008_insurance_banking_construction.pdf].

Behm M [2005]. Linking construction fatalities to the design for construction safety concept. Safety Sci 43:589–611.

Bureau of Labor Statistics [BLS] [2003–2009]. Census of fatal occupational injuries. Washington, DC: U.S. Department of Labor, Bureau of Labor Statistics [www.bls.gov/iif/oshcfoi1.htm].

BLS [2003–2009]. Current population survey. Washington, DC: U.S. Department of Labor, Bureau of Labor Statistics [www.bls.gov/cps/home.htm].

BLS [2006]. Injuries, illnesses, and fatalities in construction, 2004. By Meyer SW, Pegula SM. Washington, DC: U.S. Department of Labor, Bureau of Labor Statistics, Office of Safety, Health, and Working Conditions [www.bls.gov/opub/cwc/sh20060519ar01p1.htm].

BLS [2011]. Census of fatal occupational injuries. Washington, DC: U.S. Department of Labor, Bureau of Labor Statistics. [www.bls.gov/news.release/cfoi.t02.htm].

BLS [2011]. Injuries, illnesses, and fatalities (IIF). Washington, DC: U.S. Department of Labor, Bureau of Labor Statistics [www.bls.gov/iif/home.htm].

CFR. Code of Federal Regulations. Washington, DC: U.S. Government Printing Office, Office of the Federal Register.

CPWR [2008]. The construction chart book. 4th ed. Silver Spring, MD: Center for Construction Research and Training.

Construction Industry Research and Information Association [CIRIA] [2004]. CDM regulations: work sector guidance for designers. 2nd ed. London: Construction Industry Research and Information Association.

Driscoll TR, Harrison JE, Bradley C, Newson RS [2008]. The role of design issues in work-related fatal injury in Australia. J Safety Res 39(2):209–214.

European Foundation for the Improvement of Living and Working Conditions [1991]. From drawing board to building site (EF/88/17/FR). Dublin: European Foundation for the Improvement of Living and Working Conditions.

FindLaw [1997]. Supreme Court of Mississippi. Wanda M. Jones vs. James Reeves Contractors Inc. Case no. 93-CA-01139-SCT. March 27 [caselaw.findlaw.com/ms-supreme-court/1046041.html].

Gambatese JA, Hinze J, Haas CT [1997]. Tool to design for construction worker safety. J Arch Eng 3(1):2–41.

Haas C, O'Connor J, Tucker R, Eickmann J, Fagerlund W [2000]. Prefabrication and preassembly trends and effects on the construction workforce. Report no. 14. Austin, TX: Center for Construction Industry Studies, The University of Texas at Austin.

Hecker S, Gambatese J, Weinstein M [2005]. Designing for worker safety: moving the construction safety process upstream. Prof Saf 50(9):32–44.

Hinze J, Wiegand F [1992]. Role of designers in construction worker safety. Journal of Construction Engineering and Management 118(4):677–684.

Lipscomb HJ, Glazner JE, Bondy J, Guarini K, Lezotte D [2006]. Injuries from slips and trips in construction. Appl Ergonomics 37(3):267–274.

Luckett K [2010]. Green roof construction and maintenance. New York: McGraw Hill.

Main BW, Ward AC [1992]. What do engineers really know and do about safety? Implications for education, training, and practice. Mechanical Engineering 114(8):44–51.

New York State Department of Health [2007]. A plumber dies after the collapse of a trench wall. Case report 07NY033 [www.cdc.gov/niosh/face/pdfs/07NY033.pdf].

NIOSH [2004]. Preventing falls of workers through skylights and roof and floor openings. Cincinnati, OH: U.S. Department of Health and Human Services, Centers for Disease Control and Prevention, National Institute for Occupational Safety and Health, DHHS (NIOSH) Publication No. 2004–156.

NIOSH Fatality Assessment and Control Evaluation (FACE) Program [1991]. Construction laborer is electrocuted when crane boom contacts overhead 7200-volt power line in Kentucky. FACE9121 [www.cdc.gov/niosh/face/In-house/full9121.html].

NIOSH Fatality Assessment and Control Evaluation (FACE) Program [1983]. Fatal incident summary report: scaffold collapse involving a painter. FACE 8306 [www.cdc.gov/niosh/face/In-house/full8306.html].

NIOSH Fatality Assessment and Control Evaluation (FACE) Program [2000]. NIOSH [1999]. Driller's helper electrocuted when mast of drill rig contacted overhead power lines. Alaska FACE Investigation 99AK019 [www.cdc.gov/niosh/face/stateface/ak/99ak019.html].

NIOSH Fatality Assessment and Control Evaluation (FACE) Program [1996]. Electrician dies following a 60-foot fall through a roof—Virginia. FACE 9605 [www.cdc.gov/niosh/face/In-house/full9605.html].

NIOSH Fatality Assessment and Control Evaluation (FACE) Program [2009]. An electrical worker dies when he falls through a skylight while installing solar panels on the roof of a warehouse. California case report no. 09CA003 [www.cdc.gov/niosh/face/stateface/ca/09ca003.html].

NIOSH Fatality Assessment and Control Evaluation (FACE) Program [1999]. Laborer dies from fall through skylight while shoveling snow on roof. Wisconsin case report no. 99WI002 [www.cdc.gov/niosh/face/stateface/wi/99WI002.html].

NOHSC [2001]. CHAIR safety in design tool. New South Wales, Australia: National Occupational Health & Safety Commission.

OSHA [2001]. Standard number 1926.760: fall protection. Washington, DC: U.S. Department of Labor, Occupational Safety and Health Administration.

OSHA [ND]. Fatal Facts Accident Reports Index [foreman electrocuted]. Accident summary no. 17 [www.setonresourcecenter.com/MSDS_Hazcom/FatalFacts/index.htm].

OSHA [ND]. Fatal Facts Accident Reports Index [laborer struck by falling wall]. Accident summary no. 59 [www.setonresourcecenter.com/MSDS_Hazcom/FatalFacts/index.htm].

Rogers D [2010]. Architect fined after health and safety lapse causes death. Building Design [bdonline] 30 July [www.bdonline.co.uk/30-july-2010/20050.issue].

Szymberski R [1997]. Construction project planning. TAPPI J *80*(11):69–74.

Toole TM [2005]. Increasing engineers' role in construction safety: opportunities and barriers. Journal of Professional Issues in Engineering Education and Practice *131*(3):199–207.

Weiler S, Scholtz-Barth K [2009]. Green roof systems: a guide to the planning, design and construction of building over structure. New York: John Wiley & Sons.

USC. United States Code. Washington, DC: U.S. Government Printing Office.

Other Sources

American National Standards Institute [ANSI] [2010] ANSI/SPRI VF-1, External fire design standard for vegetative roofs [www.greenroofs.org/resources/ANSI_SPRI_VF_1_Extrernal_Fire_Design_Standard_for_Vegetative_Roofs_Jan_2010.pdf].

American Society of Civil Engineers [ASCE][www.asce.org/Content.aspx?id=7231]

Center for Urban Greenery and Ecology [CUGE] [2010]. Guidelines on design or safety on rooftop greenery. CS E02:2010. Singapore: National Parks Board.

Construction Industry Training Board [2007]. The construction (design and management) regulations 2007: industry guidance for designers. CDM 2007. Norfolk, United Kingdom: ConstructionSkills.

Construction Industry Advisory Committee [CONIAC] [www.hse.gov.uk/guidance/index.htm]

National Society of Professional Engineers [NSPE][www.nspe.org/ethics/index.html]

NIOSH Fatality Assessment and Control Evaluation Program [www.cdc.gov/niosh/face/]

NIOSH Prevention through Design program Web sites:
    www.cdc.gov/niosh/topics/PtD/
    www.cdc.gov/niosh/programs/PtDesign/

OSHA Fatal Facts Accident Reports Index:
    www.setonresourcecenter.com/MSDS_Hazcom/FatalFacts/index.htm

OSHA home page [www.osha.gov]

OSHA Anchorage Standard 29 CFR 1926.502(d)(15)

OSHA comprehensive crane standard [www.osha.gov/FedReg_osha_pdf/FED20100809.pdf]

OSHA crane regulation text [www.osha.gov/cranes-derricks/index.html]

A press release for the crane standard [www.advancedsafetyhealth.com/blog/index.php/category/cranes]

OSHA PPE publications:
    www.osha.gov/Publications/osha3151.html
    www.osha.gov/OshDoc/data_General_Facts/ppe-factsheet.pdf
    www.osha.gov/OshDoc/data_Hurricane_Facts/construction_ppe.pdf

Test Questions

1. What is the goal of Prevention through Design?

2. Give two examples of industries that have incorporated PtD into the corporate culture.

3. Name one practical benefit of PtD.

4. Give one ethical reason for PtD.

5. Give an example of a hazard associated with an urban construction site.

6. What conditions might cause the sides of an excavation to cave-in?

7. List three kinds of personal protective equipment (PPE).

8. Give three reasons why PPE is considered the solution of last resort.

9. How is PtD different from engineering controls?

10. Define "constructability."

11. Name the players who must communicate during the design phase.

12. When in the design process is the time to consider safety?

13. Why should you visit the OSHA Web site?

14. Name three construction hazards.

15. Where can you find tools to help you create safer designs?

Answers

1. The goal of PtD is to anticipate and eliminate hazards and risks at the design phase of a project/process and to make workplaces safer for workers.

2. Construction companies, computer and communications corporations, design-build contractors, electrical power providers, engineering consulting firms, oil and gas industries, water utilities

3. Accidents on the job hurt employee morale, delay project completion, and cost money.

4. Preventable accidents should be prevented! Accidents ruin lives.

5. Examples include overhead power lines, existing infrastructure (gas, electric, and sewer), pedestrians, and traffic flow.

6. A trenching accident may be caused by spring thaw, lack of shoring, cracked forms, recent precipitation, type of soil, and/or placement of heavy equipment.

7. Personal Protective Equipment, or PPE, includes items worn as a last line of defense against injury. OSHA-required PPE can include hardhats, steel-toed boots, safety glasses or safety goggles, gloves, earmuffs, full body suits, respiratory aids, face shields, and fall harnesses.

8. PPE is a solution of last resort because it
   a. requires the worker to wear it,
   b. may not fit because of limited size availability, and
   c. does not eliminate the hazard.

9. Engineering controls isolate the process or contain the hazard. PtD removes or reduces the hazard.

10. The term "constructability" implies an evaluation of a particular design in terms of cost, safety, duration, and quality. Can the design be built at a reasonable cost, within a reasonable amount of time, and result in an acceptable level of quality?

11. The entire design team must communicate, including the architect, structural engineer, civil engineer, HVAC engineer, trade representatives, and site planner.

12. Throughout!

13. OSHA regulations are updated annually. The Web site contains a summary of the latest hazard investigations. The Web site also contains information about occupational diseases.

14. Hazards include falls, tripping hazards, falling objects, loud noises, and musculoskeletal injuries.

15. Agencies such as OSHA, NIOSH, and CHAIR can provide tools to help you create safer designs.